Deterministic Numerical Modeling of Soil–Structure Interaction

Series Editor
Marc Boulon

Deterministic Numerical Modeling of Soil–Structure Interaction

Edited by

Stéphane Grange
Diana Salciarini

iSTE

WILEY

First published 2021 in Great Britain and the United States by ISTE Ltd and John Wiley & Sons, Inc.

ISTE Ltd
27-37 St George's Road
London SW19 4EU
UK

www.iste.co.uk

John Wiley & Sons, Inc.
111 River Street
Hoboken, NJ 07030
USA

www.wiley.com

Library of Congress Control Number: 2021947467

British Library Cataloguing-in-Publication Data
A CIP record for this book is available from the British Library
ISBN 978-1-78630-798-9

MIX
Paper from responsible sources
FSC® C013604

Contents

Introduction ix

Chapter 1. Hydro-mechanically Coupled Interface Finite Element for the Modeling of Soil–Structure Interactions: Application to Offshore Constructions 1

Benjamin CERFONTAINE and Robert CHARLIER

1.1. Introduction 2
1.1.1. The finite element method (FEM) 2
1.1.2. Review of existing contact formulations 3
1.1.3. Objectives 7
1.2. Governing equations of the interface problem 7
1.2.1 .Mechanical problem 8
1.2.2. Flow problem 12
1.2.3. Couplings between mechanical and flow problems 15
1.3. Numerical formulation of the element 16
1.3.1. Space and fluid pressure discretization 16
1.3.2. Mechanical problem 17
1.3.3. Flow problem 19
1.3.4. Time discretization 19
1.3.5. Stiffness matrix 20
1.4. Application 20
1.4.1. Suction caissons 20
1.4.2. Problem description 22
1.4.3. Tension loading 26
1.4.4. Lateral loading 37
1.5. Conclusion and perspectives 47
1.5.1. Conclusion 47
1.5.2. Perspectives 48
1.6. References 49

Chapter 2. DEM Approach for the Modeling of Geotechnical Structures in Interaction with Reinforcements . . . 55

Pascal VILLARD

2.1. Introduction . . . 55
2.2. Discrete modeling . . . 56
2.2.1. General concepts of the discrete modeling approach . . . 56
2.2.2. Specific interaction between discrete particles and reinforcement elements . . . 59
2.2.3. Numerical strategy for geotechnical structure modeling using DEM . . . 61
2.3. Application of the DEM to geotechnical structures in interaction with rigid piles . . . 61
2.3.1. Load transfer mechanisms within granular embankments over a network of piles . . . 64
2.3.2. Load transfer mechanisms within granular embankments over a network of piles under cyclic loadings . . . 71
2.4. Application of the DEM to geotechnical structures in interaction with flexible and deformable reinforcement – comparison with experiment results . . . 76
2.4.1. Numerical and experimental behavior of geosynthetic tubes filled with granular material . . . 78
2.4.2. Numerical and experimental behavior of granular embankments reinforced with geosynthetic in areas prone to subsidence . . . 87
2.5. Conclusion . . . 96
2.6. References . . . 96

Chapter 3. SSI Analysis in Geotechnical Engineering Problems Using a Finite Difference Method . . . 101

Daniel DIAS and Orianne JENCK

3.1. Introduction . . . 101
3.2. The finite difference method using an explicit scheme . . . 102
3.3. Application of the finite difference method to soil–structure interaction problems . . . 104
3.3.1. Structural elements . . . 105
3.3.2. Interfaces . . . 107
3.3.3. Constitutive models for soil . . . 108
3.3.4. Dimension of the problem . . . 111
3.3.5. Monotonic, quasi-static cyclic and dynamic loadings . . . 112
3.4. Some application examples in the geotechnical engineering field . . . 112
3.4.1. Reinforced retaining walls . . . 112
3.4.2. Tunneling . . . 115

3.4.3. Soft soil improvement using vertical rigid piles 121
3.5. Conclusion 138
3.6. References 139

Chapter 4. Macroelements for Soil–Structure Interaction 143

Diana SALCIARINI, Stéphane GRANGE, Claudio TAMAGNINI and Panagiotis KOTRONIS

4.1. Introduction 143
4.2. The concept of generalized forces: Eurocode 8 recommendations. 145
4.3. Macroelements for shallow foundations 148
4.3.1. Generalities 148
4.3.2. Macroelements formulated in the framework of hardening elastoplasticity 150
4.3.3. Macroelements formulated in the framework of hypoplasticity 151
4.4. The considered macroelements 154
4.4.1. The elastoplastic macroelement 154
4.4.2. The hypoplastic macroelement. 156
4.5. Case study: seismic response of a reinforced concrete viaduct 160
4.5.1. Features of the viaduct 160
4.5.2. The finite element model of the viaduct and its foundations 161
4.5.3. Seismic input 163
4.6. Calibration of the macroelements 164
4.7. Results of the numerical simulations 167
4.7.1. Forces and displacements in the structural elements 167
4.7.2. Displacements of the abutment and the foundations 170
4.8. Concluding remarks 172
4.9. References 175

Chapter 5. Urban Seismology: Experimental Approach to Soil–Structure Interaction Towards the Concept of Meta-city 181

Philippe GUÉGUEN, Philippe ROUX and Andrea COLOMBI

5.1. Introduction 181
5.1.1.Observation of soil–structure interaction under weak and strong seismic loading 183
5.1.2. Contamination of urban seismic motion by the vibration of buildings 196
5.1.3. Conclusion 202
5.2. References 203

List of Authors 211

Index. 213

Introduction

The deterministic numerical modeling of Soil–Structure Interaction (SSI) uses the tools of Mechanics, and of their transposition, according to the methods of the numerical discretization. An SSI analysis should take into account the differences between the properties of the soil and the structure, possibly leading to local relative displacements. In addition, the presence of water enforces to consider the interaction on the hydro-mechanical framework, involving pore pressures and fluid flow.

Non-deterministic methods, making use of artificial intelligence and progressive learning, are not considered here.

The geotechnical works are usually designed referring to the following three situations:

1) The *pseudo-static case*: where the loadings are nearly time independent (a small number of cycles are accepted).

2) The *cyclic case*: involving a large number of slow cycles (slow versus the time scales of any physical phenomena involved).

3) The *dynamic (mostly seismic) case*: where inertia forces are taken into account.

In the pseudo-static and cyclic cases, the SSI is defined from a local point of view, whereas in dynamic conditions it is examined from a more global angle. The soil–structure interfaces, transfer zones of the loads acting between soil and structure, and often zones of large localized deformations have a significant importance. Special strategies and experimental characterization tests have been developed for describing them.

This volume gives an overview on the main modeling methods developed in geotechnical engineering in order to describe the SSI in various situations:

– The classical and well-known finite element method (FEM) using interface or contact elements available for coupled hydro-mechanical problems and in which the local plastic energy dissipation contributes to the classical global damping.

– The distinct element method (DEM) in which the contact zones between soil and structure are described by local adapted interaction laws.

– The finite difference method (FDM) using explicit algorithm in which interface or contact elements are available and straightforward for coupled hydro-mechanical, highly nonlinear with high deformation and dynamic problems.

– The more recent approaches based on the *macro-element* concept and generalized variables in an incremental form available in static and dynamic situations, where a large volume of soil underlying and surrounding the-structure is represented by a small number of degrees of freedom. This method could be considered as an improved extension of the stiffness coefficient methods known for the piles as the t-z and p-y curves.

– The seismology approach with inertia and viscous equivalent damping forces, in which the SSI is developed between several structures (like buildings in a town) using the notion of meta-materials. This point of view is especially useful when analyzing the movement of the buildings induced by a seismic event at the scale of a town.

1

Hydro-mechanically Coupled Interface Finite Element for the Modeling of Soil–Structure Interactions: Application to Offshore Constructions

This chapter presents how the multiphysical soil–structure interactions can be modeled using the finite element method (FEM). This method takes into account different nonlinearities, due to geometry, material behavior, physical couplings and interfaces. The FEM allows the volume to be discretized into a finite number of elements and a finite number of degrees of freedom. The numerical techniques for solving a nonlinear problem in the framework of the FEM have been extensively studied and are now very robust.

The behavior of offshore foundations is highly dependent on interface properties. Upon large overturning moment or lateral load, the soil–structure interaction results in rotation, sliding or gap opening. Hydro-mechanical couplings have become particularly important in many applications, as the loading rate and soil permeability are such that the soil behavior is not completely drained or undrained.

A hydro-mechanically coupled interface element is proposed in this work to simulate the behavior of suction caissons embedded in sand. This interface element simulates the mechanical problem with a penalty method and reproduces the friction at the interface as a simple Coulomb model. The interface element introduces hydro-mechanical couplings based on the gap opening (storage, longitudinal permeability evolution) and the definition of an effective contact stress.

Chapter written by Benjamin CERFONTAINE and Robert CHARLIER.

1.1. Introduction

The advent of numerical methods and, especially, finite element approaches has provided engineers with formidable tools to predict the behavior of constructions. Most of the time, the soil can be considered as a homogeneous medium behaving in a purely drained or undrained manner. However, in some cases, depending on the soil permeability and the loading rate, partially drained behaviors should be considered. In this case, hydro-mechanical couplings must be included in the finite element formulation for both volume and interface elements.

The development of renewable energy sources is the greatest challenge of the 21st Century. A large number of offshore wind farms have been developed throughout the last decade, especially in shallow waters in the North Sea where a sandy seabed is mostly encountered. The offshore wind industry is expected to grow exponentially, followed by the development of more recent wave and tidal energy devices. Whether they are bottom-fixed or floating, their foundations or anchors will have to be designed. The finite element simulation of anchors and foundations has become more popular among offshore wind farm designers. However, the modeling of the interface between the soil and the foundations is still very rudimentary.

From the mechanical point of view, shearing along the interfaces participates in the strength of the foundation against applied loads in any direction (horizontal, vertical or moment). When the maximum capacity is overcome, sliding occurs between the soil and the foundation. Each movement of the foundation results in a fluid flow, since it lies under the sea. This affects the interface behavior by modifying the maximum shear stress available or by inducing a suction effect, because of a gap opening (e.g. for suction caissons). The finite element modeling of these complex interactions at the interface is the topic of this chapter.

1.1.1. *The finite element method (FEM)*

The finite element method (FEM) is one of the most popular numerical methods for solving partial differential equations for boundary value problems [ZIE 00]. The main concept of the method is to decompose the entire continuum domain to be modeled into a collection of subdomains, called elements. The continuum field of any physical variable of interest (e.g. displacement, pore water pressure) is approximated over each element, based on a finite number of variables (physical unknowns) and pre-defined interpolation functions. The partial differential equations that must be solved over the entire continuum domain can then be approximated and discretized for each element. The set of discrete equations associated with each element can be combined into a single global system and solved algebraically. The solution of this set of equations is the one that minimizes the error of the approximation with respect to the actual solution. The FEM has many advantages, such as the ability to simulate complex geometries, incorporate different constitutive laws, involve multiphysical couplings and analyze stress and internal variables locally.

1.1.2. *Review of existing contact formulations*

Numerically solving the mechanical contact problem is not a recent topic, and several books have been dedicated to this issue [JOH 92, WRI 06]. Within the framework of finite element methods, two general approaches exist in order to manage contact between two solid bodies, namely the thin layer and the zero-thickness approach, as shown in Figure 1.1.

The first approach consists of explicitly modeling the contact zone with special finite elements, designed to encounter large shear or compression deformation [DES 84, SHA 93, WRI 13, WEI 15]. The second approach, adopted in the following, involves special boundary elements which have no thickness, namely zero-thickness elements [GOO 68, CHA 88, DAY 94, HAB 98, WRI 06]. They discretize a potential zone of contact. A gap between each side of the interface and a probable other solid is computed at each time step to detect contact. Three main ingredients are necessary to formulate this approach:

– a criterion to rule the contact detection/loss and contact pressure evolution;

– a constitutive law to describe the shear/normal behavior(s);

– a technique to discretize the contact area between solids and to compute the gap function g_N, namely the distance between two solids.

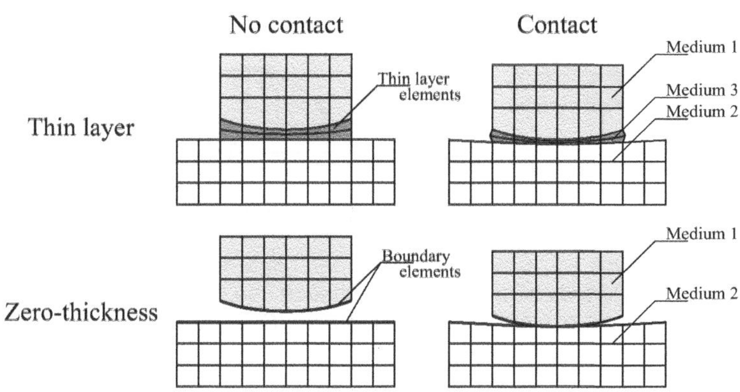

Figure 1.1. *Comparison between the thin layer and zero-thickness approaches in the case of Hertzian contact (source: [CER 15])*

The normal *contact constraint* ensures that two perfectly smooth solids in contact do not overlap each other. This mathematical criterion relates the gap function g_N and the contact pressure p_N. It states that when the gap is null, the solids are in contact and a contact pressure prevents their overlapping. When the pressure is equal to zero and the gap is greater than or equal to zero, there is no contact.

The relationship between gap and pressure evolution can be termed as *high precision* or *low precision*. In the first case, this relationship is physically based and the pressure increases with a gap that can be negative, leading to interpenetration ($g_N < 0$). This is especially true in rock mechanics, where surfaces in contact are not perfectly smooth and the normal pressure increases with deformation of asperities [GEN 90].

A low-precision contact is ensured on a purely geometrical basis. Two surfaces of bodies in contact are considered to be perfectly smooth and interpenetration is theoretically not allowed ($g_N \leq 0$) [WRI 04]. There are only two states: in contact or not in contact. The *Lagrange multiplier* method exactly ensures this condition [BEL 91, WRI 06]. Lagrange multipliers are introduced to enforce the constraint, and correspond to the contact pressures. On the other hand, the penalty method [HAB 98] regularizes the constraint by authorizing a limited interpenetration of the solids in contact. The contact pressure is linearly proportional to this interpenetration by a coefficient called the penalty coefficient. Both methods are compared in Figure 1.2.

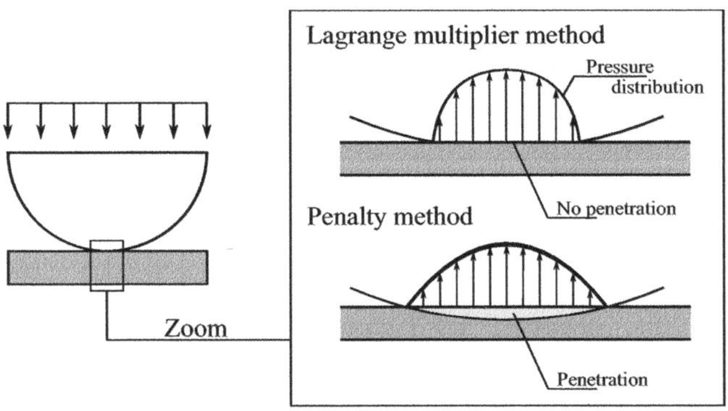

Figure 1.2. *Comparison between Lagrange multiplier and penalty methods on deformation and distribution of contact pressures (source: [CER 15])*

The penalty method is adopted in the following for its simplicity, since it does not require us to introduce a variable number of unknowns (Lagrange multipliers) at each step. Both solutions are theoretically identical if the penalty coefficient tends to infinity. However, increasing this coefficient too much creates a badly conditioned stiffness matrix, which becomes difficult to solve.

The mobilization of shear stress along the interface is very important in many applications. The maximum shear stress that could be mobilized in a tangential plane is strongly dependent on the normal pressure. The most basic relationship between them is the classical Coulomb criterion. However, similar to the normal behavior, more complex constitutive laws may be defined: for example, to describe rock joints [ALO 13, ZAN 13] or soil–structure interfaces [LIU 06, STU 16], including critical state, we use dilative normal behavior and even cyclic degradation [LIU 08].

The normal contact constraint is defined as a continuous condition over the boundaries of the solids in contact. It must be discretized in the finite element framework. Computation of the gap (g_N) and enforcement of the contact constraint are different depending on the method, as shown in Figure 1.3.

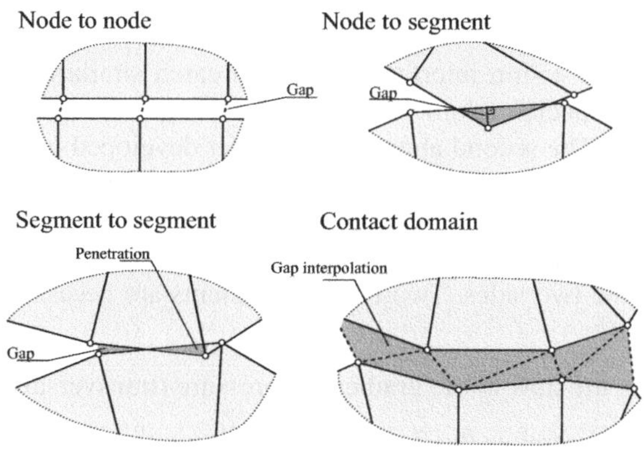

Figure 1.3. *Comparison between the discretization methods of the contact area (source: [CER 15])*

The simplest method, namely *node-to-node* discretization, imposes the constraint on a pair of nodes [KLA 88], limiting the problems to small relative tangential displacement. The *node-to-segment* discretization overcomes this drawback [HAL 85]. The gap function is computed by the projection of a node of one side, termed the *slave surface*, on a segment of the other side, termed the *master surface*. This method is sensitive to sudden changes in the projection direction, but smoothing techniques strongly reduce this issue [WRI 01]. The *contact domain* discretization does not involve any projection method [OLI 09, HAR 09]. The gap between the solids potentially in contact is discretized by a fictitious mesh. Therefore, the gap function is continuous between them and avoids many discrepancies and loss of unicity due to projection.

The *segment-to-segment* discretization [HAB 98, PUS 04, FIS 06] is based on the mortar method developed in [BEL 98]. In this case, the contact constraint is applied over the element in a weak sense. The gap function is computed by the closest-point projection of a point of the non-mortar surface onto the mortar surface, which is given more importance. In the following, this method is adopted. The distance is computed at each integration point of the element and extrapolated over it by means of interpolation functions.

In many examples, interfaces may be considered as preferential paths for fluid flows. Therefore, a description of a longitudinal fluid flow, namely parallel to the interface, must be defined. Boussinesq [BOU 68] first provided a mathematical law that characterized the laminar flow of a viscous incompressible fluid between two smooth parallel plates. The total fluid flow was proved to be proportional to the cube of the aperture between the plates, and this relationship was termed the *cubic law*. The applicability of this law to rock mechanics was proven [WIT 80, TSA 81, ORO 98] under some hypotheses [OLS 01].

The flow problem within interfaces may be treated similarly to the mechanical contact problem by special volume elements [THI 14] or the zero-thickness method [GUI 02, SEG 08a]. The second approach is further developed in the following to be consistent with the mechanical formulation. The flow problem is intrinsically coupled with the mechanical problem through the opening/closing of the gap. In addition to the flow along the interface direction, transversal flow also occurs, causing a drop of pressure between the two sides. Two basic ingredients are necessary to describe this part:

– a law relating the flow to the gradient of pressure (transversal/longitudinal);

– a technique to discretize the flow within and through the interface.

The *single-node* discretization of flow is the simplest one, as shown in Figure 1.4. It simply superposes a discontinuity for fluid flow into a continuous porous medium [BER 88]. In this case, there is no hydro-mechanical coupling and the opening of the discontinuity is constant and user-defined. It acts as a pipe.

The *double-node* discretization considers the longitudinal gradients of pressure on each side of the interface. The longitudinal flow depends on the average gradient. The transversal flow is a function of the transversal difference between the pressure and a parameter called transversal transmissivity [NG 97, GUI 02, SEG 08a, SEG 08b]. In this case, the mechanical and flow behaviors of the interface are able to coexist and are coupled.

The last option dissociates the discretization of the longitudinal flow from both sides of the interface by adding nodes inside the interface. The nodes have only a single pressure degree of freedom. This method is termed *triple-node* discretization

[GUI 02, JHA 14, CER 15]. The underlying hypothesis is that the field of pressure is homogeneous inside the interface. In this case, there are two drops of pressure between each side of the interface and inside the interface.

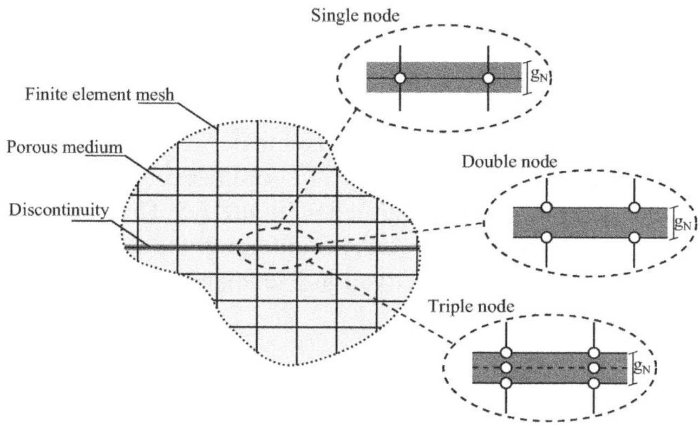

Figure 1.4. *Comparison between the discretization methods of the flow within and through the interface (source: [CER 15])*

1.1.3. *Objectives*

The first objective of this chapter is to provide a basic formulation of hydro-mechanically coupled finite elements of the interface. The developments are sufficiently general to be applied to any finite element code. In addition, any constitutive law, from the most simple to the most complicated, can be inserted into this formulation. The governing equations and finite element formulations are provided in sections 1.2 and 1.3.

The second objective is to highlight the importance of considering hydro-mechanically coupled interfaces in the soil–structure interaction. An example of suction caisson modeling is provided in section 1.4 to illustrate the inherently partially drained (between drained and undrained) behavior of this foundation in the offshore environment and how interface elements are used to simulate this behavior.

1.2. Governing equations of the interface problem

The interface finite element is modeled using the zero-thickness approach, which does not require any remeshing technique. The contact constraint is enforced by a penalty method that is easy to implement and does not require any additional unknowns. The mechanical constitutive law ruling the shear behavior is the classical Coulomb model. The fluid flow problem is solved using a three-node discretization. The couplings arise from the variation of storage and permeability with gap opening, as well as the definition of an effective stress. The purpose of this section is to derive the main equations of the problem and to formulate the continuum interface problem.

1.2.1. *Mechanical problem*

1.2.1.1. *Definition of the mechanical problem and gap function*

Let us consider two deformable porous media Ω^1 and Ω^2 in their current configurations at time t. The global system of coordinates is termed $(\mathrm{E}_1, \mathrm{E}_2, \mathrm{E}_3)$. A 2D cross-section of these bodies is illustrated in Figure 1.5. Their evolution is assumed to be quasi-static. Their boundaries in current configurations are denoted by Γ^1 and Γ^2. Imposed displacement (Dirichlet) and traction (Neumann) boundaries are respectively denoted by $\Gamma^i_{\bar{u}}$ and $\Gamma^i_{\bar{t}}$.

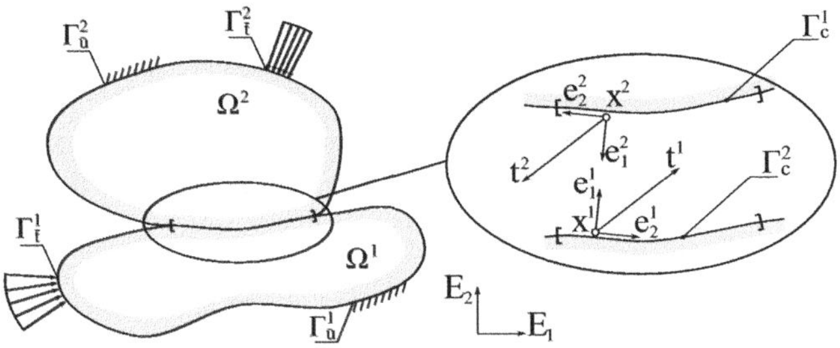

Figure 1.5. *Statement of the mechanical problem and cross-section of the 3D problem in the* $(\mathrm{E}_1, \mathrm{E}_2)$ *plane (source: [CER 15])*

Γ^1_c and Γ^2_c denote both parts of the boundary where contact is likely to occur. In this area, a local system of coordinates $(\mathbf{e}^1_1, \mathbf{e}^1_2, \mathbf{e}^1_3)$ is defined along the mortar side Γ^1_c, as shown in Figure 1.5, where $\mathbf{e}^1_1$ denotes the normal to the surface. The closest-point projection $\bar{\mathbf{x}}^1$ of a point of $\mathbf{x}^2$ of the boundary Γ^2_c onto Γ^1_c is defined such that [WRI 06]

$$g_N = \left(\mathbf{x}^2 - \bar{\mathbf{x}}^1\right) \cdot \bar{\mathbf{e}}^1_1, \qquad [1.1]$$

where $(\bar{\mathbf{e}}^1_1, \bar{\mathbf{e}}^1_2, \bar{\mathbf{e}}^1_3)$ denotes the local system of coordinates at point $\bar{\mathbf{x}}^1$. This function g_N is referred to as the gap function, where the subscript N stands for the normal direction. If there is no contact between the solids, g_N is positive. The contact is termed *ideal* if there is no interpenetration of the solids. This can be enforced if the Lagrange multiplier method is used. However, if the penalty method is used, interpenetration (negative gap function) is necessary to generate the contact pressure.

The definition of a relative tangential displacement between two points in the plane of contact has no meaning in the field of large displacement [HAB 98]. Rather, normal

(*N*) and tangential (*T1* and *T2*) velocities are defined in the local system of coordinates. They are gathered into the vector $\dot{\mathbf{g}}$ such that

$$\dot{\mathbf{g}} = \dot{g}_N \, \mathbf{e}_1^1 + \dot{g}_{T1} \, \mathbf{e}_2^1 + \dot{g}_{T2} \, \mathbf{e}_3^1. \quad [1.2]$$

1.2.1.2. *Normal contact constraint*

The contact between two solids gives rise to non-zero stress vectors $\mathbf{t}^1 = -\mathbf{t}^2$ along their common boundary, as shown in Figure 1.5. These vectors are described in the corresponding local system of coordinates at each contact point such that

$$\mathbf{t}^1 = -\mathrm{p}_N \, \mathbf{e}_1^1 + \tau_1 \, \mathbf{e}_2^1 + \tau_2 \, \mathbf{e}_3^1, \quad [1.3]$$

where p_N is the normal pressure, and τ_1 and τ_2 are the shear stresses in both directions in the plane of the interface. The ideal contact constraint is summarized into the Hertz–Signorini–Moreau condition [WRI 06]:

$$\mathrm{g}_N \geq 0, \quad \mathrm{p}_N \geq 0 \quad \text{and} \quad \mathrm{p}_N \, \mathrm{g}_N = 0. \quad [1.4]$$

If there is no contact, the gap function g_N is positive and the contact pressure p_N is null. In case of contact, the gap function is null and the contact pressure is positive.

This condition is no longer verified if the penalty method is used. In case of contact, the relation between the pressure and the gap function can be written as

$$\dot{\mathrm{p}}_N = -\mathrm{K}_N \, \dot{\mathrm{g}}_N, \quad [1.5]$$

where the minus sign ensures that the contact pressure is positive when inter-penetration increases, i.e. $\mathrm{g}_N < 0$ and $\dot{\mathrm{g}}_N < 0$.

1.2.1.3. *Tangential contact constraint*

When solids are in contact, the ideal tangential behavior of the interface distinguishes between the *stick* and *slip* states [WRI 06]. In the former state, two points in contact cannot move tangentially, i.e. they remain stuck together during the simulation, i.e. $\dot{\mathrm{g}}_{T1} = 0$ and $\dot{\mathrm{g}}_{T2} = 0$. The second condition involves a relative tangential displacement in the plane of the interface. This is summarized in a condition similar to equation [1.4]:

$$\dot{\mathrm{g}}_{Ti}^{sl} \geq 0, \quad f(\mathbf{t}, \mathbf{q}) \leq 0 \quad \text{and} \quad \dot{\mathrm{g}}_{Ti}^{sl} \, f(\mathbf{t}, \mathbf{q}) = 0 \qquad i = 1, 2 \quad [1.6]$$

where $\dot{\mathrm{g}}_{Ti}^{sl}$ is the variation of the non-recoverable displacement in each tangential direction. It is related to the variation of tangential displacement

$$\dot{\mathbf{g}}_T = \mathrm{sign}(\dot{\tau}_1) \, \dot{\mathrm{g}}_{T1}^{sl} \, \mathbf{e}_2^1 + \mathrm{sign}(\dot{\tau}_2) \, \dot{\mathrm{g}}_{T2}^{sl} \, \mathbf{e}_3^1. \quad [1.7]$$

The *stick* and *slip* states are distinguished by the criterion $f(\mathbf{t}, \mathbf{q})$. It depends on the stress state $\mathbf{t}$ and a set of internal variables $\mathbf{q}$. The evolution of the stress state within the interface depends on the constitutive law described hereafter.

The ideal *stick* state, $\dot{\mathbf{g}}_T = 0$, is also regularized by the penalty method, i.e. a relative displacement is allowed. Therefore, the relation between the shear stress and the tangential variation of displacement is given by

$$\dot{\tau}_i = \mathrm{K}_T \, \dot{\mathrm{g}}_{Ti} \qquad i = 1, 2. \qquad [1.8]$$

1.2.1.4. *Constitutive law*

The constitutive law adopted is a simple Coulomb model despite the actual behavior of the soil–structure interface, and rock joints are proved to be more complex than this simple criterion [HU 04, DEJ 06, HO 11, ZAN 13]. However, the constitutive law is kept as basic as possible, since the examples presented are complex enough. On the other hand, any nonlinear relation between stress and strain increments can be substituted into the equations presented below. The interested reader should refer to [SHA 97, MOR 02, LIU 08, STU 16] for a deeper insight into more complex constitutive laws.

The constitutive law depends only on the stress state $\mathbf{t}$ within the interface and a single parameter, the friction coefficient μ. Mathematically, it can be written as

$$f(\mathbf{t}, \mu) = \underbrace{\sqrt{(\tau_1)^2 + (\tau_2)^2}}_{\|\tau\|} - \mu \, \mathrm{p}_N. \qquad [1.9]$$

where $\|\tau\|$ is the norm of the tangential stresses. The criterion is presented in Figure 1.6. In the absence of contact, the stress state lies on the apex of the criterion. Both normal pressure and tangential stresses are null, i.e. $\mathbf{t} = \mathbf{0}$. If the combination of tangential and normal stresses lies below the criterion ($f < 0$), the tangential state is considered as *stick*. Otherwise, if the stress state lies on the criterion ($f = 0$), the tangential state is considered as *sliding*.

The tangential behavior is regularized by two shear penalty coefficients. Therefore, there is no ideal *stick state* but, rather, a slight tangential displacement occurs. The criterion may be interpreted as an elastoplastic relation and the shear penalty coefficients act as shear moduli, acquiring a physical meaning. Hence, the incremental relation between variations of stresses $\dot{\mathbf{t}}$ and variations of the gap function $\dot{\mathbf{g}}$ can be written as

$$\begin{bmatrix} \dot{\mathrm{p}}_N \\ \dot{\tau}_1 \\ \dot{\tau}_2 \end{bmatrix} = \underbrace{\begin{bmatrix} -K_N & 0 & 0 \\ 0 & \mathrm{K}_T & 0 \\ 0 & 0 & \mathrm{K}_T \end{bmatrix}}_{\mathbf{D}^e} \cdot \begin{bmatrix} \dot{\mathrm{g}}_N \\ \dot{\mathrm{g}}_{T,1} \\ \dot{\mathrm{g}}_{T,2} \end{bmatrix}, \qquad [1.10]$$

where $\mathbf{D}^e$ is equivalent to the elastic compliance tensor. When the interface reaches the slip state, an elastoplastic compliance tensor $\mathbf{D}^{ep}$ is defined such that

$$\begin{bmatrix} \dot{\mathrm{p}}_N \\ \dot{\tau}_1 \\ \dot{\tau}_2 \end{bmatrix} = \underbrace{\begin{bmatrix} -K_N & 0 & 0 \\ -\mu\,\mathrm{K}_N \dfrac{\tau_1}{\|\tau\|} & \mathrm{K}_T \left(1 - \dfrac{(\tau_1)^2}{\|\tau\|^2}\right) & -\mathrm{K}_T \dfrac{\tau_1\,\tau_2}{\|\tau\|^2} \\ -\mu\,\mathrm{K}_N \dfrac{\tau_2}{\|\tau\|} & -\mathrm{K}_T \dfrac{\tau_1\,\tau_2}{\|\tau\|^2} & \mathrm{K}_T \left(1 - \dfrac{(\tau_2)^2}{\|\tau\|^2}\right) \end{bmatrix}}_{\mathbf{D}^{ep}} \cdot \begin{bmatrix} \dot{\mathrm{g}}_N \\ \dot{\mathrm{g}}_{T,1} \\ \dot{\mathrm{g}}_{T,2} \end{bmatrix}. \qquad [1.11]$$

This tensor was introduced in [CHA 88], which is based on a non-associated flow rule.

	No contact	Stick	Slip
p_N	$= 0$	> 0	> 0
$\|\tau\|$	$= 0$	≥ 0	$= \mu \cdot p_N$

Table 1.1. *Stress state in the interface*

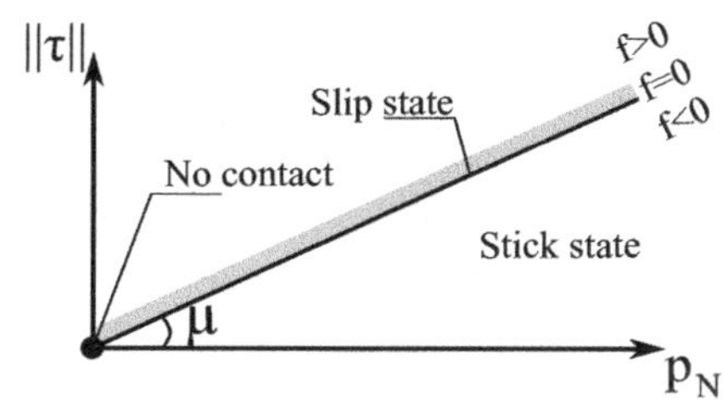

Figure 1.6. *Mohr–Coulomb criterion (source: [CER 15])*

1.2.1.5. *Continuum formulation*

Each solid Ω^i verifies the classic mechanical equilibrium equations in quasi-static conditions [BOR 13]. Solving the mechanical contact problem consists of finding the field of displacement $\mathbf{u}$ for all points $\mathbf{x} \in \Omega^i$ to verify these equations, subjected to the contact constraint equations [1.4] and [1.6].

Considering a field of admissible virtual velocities $\delta\dot{\mathbf{x}}$ on Ω^i, the weak form of the principle of virtual power can be written as

$$\sum_{i=1}^{2} \left[\int_{\Omega^i} \boldsymbol{\sigma} : \boldsymbol{\epsilon}\left(\delta\dot{\mathbf{x}}\right)\, \mathrm{d}\Omega \right] = \sum_{i=1}^{2} \left[\int_{\Omega^i} \bar{\mathbf{f}} : \delta\dot{\mathbf{x}}\, \mathrm{d}\Omega + \int_{\Gamma^i_{\bar{\mathbf{t}}}} \bar{\mathbf{t}} : \delta\dot{\mathbf{x}}\, \mathrm{d}\Omega + \int_{\Gamma^i_c} \mathbf{T}^i : \delta\dot{\mathbf{x}}\, \mathrm{d}\Gamma \right], \qquad [1.12]$$

where $\bar{\mathbf{f}}$ represents the body forces, $\bar{\mathbf{u}}$ represents the imposed displacements, $\bar{\mathbf{t}}$ represents the imposed tractions, $\mathbf{n}$ is the normal to $\Gamma^i_{\bar{t}}$ and $\mathbf{T}^i$ is the projection of the local stress tensor $\mathbf{t}^i$ in global coordinates. The equality of equation [1.12] is enforced when the contact area Γ^i_c is known.

1.2.2. *Flow problem*

1.2.2.1. *Definition of the problem*

Let us consider a discontinuity of very thin width embedded in a porous medium in its current configuration, as shown in Figure 1.7. This could represent, for example, an open fault within a rock mass or a gap between a foundation and the surrounding soil. This discontinuity creates a preferential path for fluid flow. Moreover, there is a transversal fluid flow between the rock mass and the discontinuity.

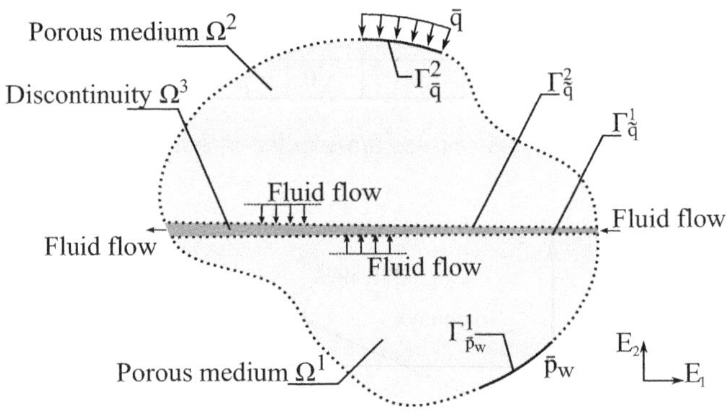

Figure 1.7. *Definition of the flow problem (cross-section of the 3D case in the* (E_1, E_2) *plane), porous medium, discontinuity and boundaries (source: [CER 15])*

There is a conceptual difference between the treatment of the mechanical and flow contact problems. The mechanical contact constraint consists of a non-zero pressure p_N applied along the contact zone Γ_c between the two solids Ω^1 and Ω^2.

On the contrary, the opening of the discontinuity creates a gap g_N filled with a fluid. This gap creates a new volume Ω^3 in which fluid flow takes place, as shown in Figure 1.8. It is bounded by the two porous media Ω^1 and Ω^2. Their boundary is termed $\Gamma^1_{\tilde{q}}$ and $\Gamma^2_{\tilde{q}}$. Therefore, $\Gamma_{\tilde{q}}$ represents a boundary where the solids are close enough, fluid interactions take place and mechanical contact is likely to occur. It always includes the contact zone Γ_c.

Ω^3 is modeled as an equivalent porous medium. The fluid flow within it is described by the cubic law. Fluid flows exist between the inner volume Ω^3 and both adjacent porous media Ω^1 and Ω^2. This flow is a function of the difference of pressure between them. This is a non-classical boundary condition since it is not an imposed flux or an imposed pressure.

Finally, imposed flux and pressure boundaries on Ω^1 and Ω^2 are respectively denoted by $\Gamma^i_{\bar{\mathrm{q}}}$ and $\Gamma^i_{\bar{\mathrm{p}}_w}$.

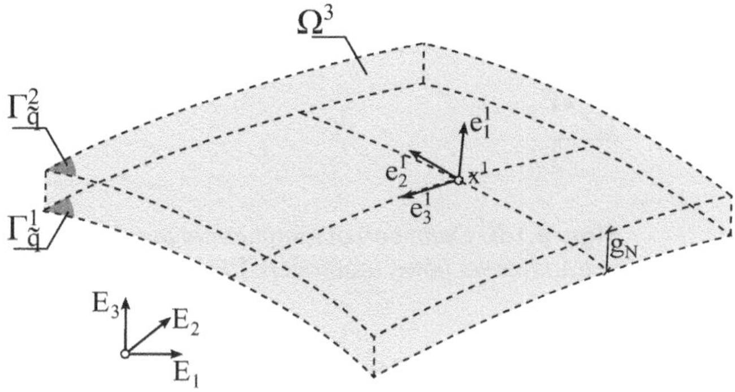

Figure 1.8. *Definition of the equivalent interior porous medium Ω^3 bounded by $\Gamma^1_{\bar{q}}$ and $\Gamma^2_{\bar{q}}$ (source: [CER 15])*

1.2.2.2. *Fluid flow formulation*

A three-node formulation is adopted to describe the fluid flow through and within the interface, as described in Figure 1.9. Therefore, fluid pressures on each side of the interface (p_{w1} and p_{w2}) and the inner fluid pressure (p_{w3}) are the fluid variables. At each point within the interface, four fluxes are defined:

– two longitudinal fluxes (f_{wl1} and f_{wl2}) in the local tangential directions ($\mathbf{e}^1_2, \mathbf{e}^1_3$) in the plane of the interface;

– two transversal fluxes (f_{wt1} and f_{wt2}) in the local normal direction ($\mathbf{e}^1_1$).

The generalized Darcy's law is assumed to reproduce the local longitudinal fluid flows f_{wl1} and f_{wl2} in the plane of the interface. It reads, in each local tangential direction ($\mathbf{e}^1_2, \mathbf{e}^1_3$):

$$\mathrm{f}_{wl(i-1)} = -\frac{k_l}{\mu_w} \left(\nabla_{\mathbf{e}^1_i} \mathrm{p}_{w3} + \rho_w \, \mathrm{g} \, \nabla_{\mathbf{e}^1_i} z \right) \rho_w \quad \text{for} \quad i = 2, 3 \qquad [1.13]$$

where $\nabla_{\mathbf{e}^1_i}$ is the gradient in the direction $\mathbf{e}^1_i$, μ_w is the dynamic viscosity of the fluid, g is the acceleration of gravity, ρ_w is the density of the fluid and k_l is the permeability. These equations make the assumption that the fluid pressure is constant over the gap g_N and the gap is fully saturated.

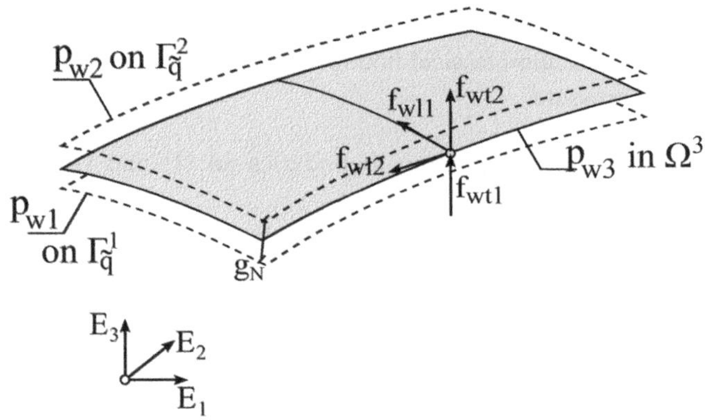

Figure 1.9. *Definition of longitudinal and transversal flows (source: [CER 15])*

Each transversal fluid flux is a function of a transversal conductivity T_{wi} and the drop of pressure across $\Gamma^i_{\tilde{q}}$. They can be written as

$$f_{wt1} = \rho_w \, T_{w1} \, (p_{w1} - p_{w3}) \qquad \text{on} \quad \Gamma^1_{\tilde{q}}, \qquad [1.14]$$

$$f_{wt2} = \rho_w \, T_{w2} \, (p_{w3} - p_{w2}) \qquad \text{on} \quad \Gamma^2_{\tilde{q}}. \qquad [1.15]$$

1.2.2.3. *Continuum formulation*

The two actual porous media (Ω_1 and Ω_2) and the equivalent porous medium Ω_3 verify the classic hydraulic equilibrium equations [LEW 98]. Solving the contact problem consists of finding the pore water distribution on Ω^i to verify the equilibrium equations and satisfy the non-classical boundary conditions [1.14]–[1.15] over $\Gamma^i_{\tilde{q}}$. Considering a field of admissible virtual pore water pressures δp_w on Ω, the weak formulation of the virtual power principle can be written as

$$\sum_{i=1}^{3} \left[\int_{\Omega^i} \dot{S} \, \delta p_w - \mathbf{f}_w \cdot \boldsymbol{\nabla} (\delta p_w) \, d\Omega \right] = \sum_{i=1}^{3} \left[\int_{\Omega^i} \bar{Q} \, \delta p_w \, d\Omega + \int_{\Gamma^i_{\bar{q}}} \bar{q} \, \delta p_w \, d\Gamma + \int_{\Gamma^i_{\tilde{q}}} \tilde{q} \, \delta p_w \, d\Gamma \right] \qquad [1.16]$$

where $\mathbf{f}_w$ is the fluid flux at point $\mathbf{x}$, $\dot{S}$ is the storage term, $\bar{Q}$ is the imposed volume source, $\bar{p}_w$ is the imposed fluid pressure, $i = 1, 2$ corresponds to the two porous media in contact and $i = 3$ to the volume of the interface. The fluid flow $\tilde{q}$ along the boundary corresponds to the transversal fluid flows f_{wti} defined in equations [1.14] and [1.15]. The source term $\bar{Q}$ associated with Ω^3 is null.

The mechanical problem gives more importance to the mortar side Γ_c^1. Similarly, the integral over Ω^3 is transformed into a surface integral over $\Gamma_{\tilde{q}}^1$. This hypothesis is valid because it is assumed that the inner pressure is constant over the aperture g_N of the interface. Therefore, equation [1.16] for $i = 3$ can be finally written as

$$\int_{\Gamma_{\tilde{q}}^1} \left[\dot{S}\,\delta p_w - f_{wl1}\,\nabla_{\mathbf{e}_2^1}(\delta p_w) - f_{wl2}\,\nabla_{\mathbf{e}_3^1}(\delta p_w)\right] g_N \,d\Gamma =$$

$$\int_{\Gamma_{\tilde{q}}^1} \rho_w\,T_{w1}\,(p_{w1} - p_{w3})\,\delta p_w - \rho_w\,T_{w2}\,(p_{w3} - p_{w2})\,\delta p_w\,d\Gamma, \qquad [1.17]$$

where $\nabla_{\mathbf{e}_i^1}$ is the gradient in the $\mathbf{e}_i^1$ direction.

In the porous media Ω^1 and Ω^2, the storage component $\dot{S}$ is coupled with the deformation of the solid skeleton. The treatment of this component for Ω^3 is different and detailed hereafter.

1.2.3. *Couplings between mechanical and flow problems*

The couplings between mechanical and flow problems depend strongly on the gap function g_N. The first coupling arises from the definition of the cubic law with respect to the gap opening (hydraulic and mechanical openings are treated as an identical variable). Therefore, if contact exists, the gap is negative and permeability is equal to zero. However, from a physical point of view, there could be a path for fluid flow through asperities of a rough surface (residual opening). Moreover, from the numerical point of view, a null permeability may lead to a badly conditioned problem. Therefore, a very low residual opening D_0 is added. Hence, the permeability is computed according to [OLS 01, GUI 02]

$$k_l = \begin{cases} \dfrac{(D_0)^2}{12} & \text{if} \quad g_N \leq 0 \\ \dfrac{(D_0 + g_N)^2}{12} & \text{otherwise.} \end{cases} \qquad [1.18]$$

It is updated during the simulation to take into account the possible gap aperture.

A second coupling is created by the storage component $\dot{S}$. The variation of the total mass of fluid $\dot{M}_f$ stored in Ω^3 comes respectively from the variation of the fluid density, the opening/closing of the gap and the variation of the surface of the discontinuity, namely

$$\dot{M}_f = \underbrace{\left(\dot{\rho}_w\,g_N + \rho_w\,\dot{g}_N + \rho_w\,g_n\,\frac{\dot{\Gamma}_{\tilde{q}}}{\Gamma_{\tilde{q}}}\right)}_{\dot{S}}\,\Gamma_{\tilde{q}}. \qquad [1.19]$$

where $\dot{\mathrm{S}}$ is the storage term of equation [1.16]. In the following, the fluid is assumed to be incompressible $\dot{\rho}_w = 0$, and only the geometrical storage is taken into account. In many applications, the main component of the storage is due to the opening/closing of the interface $\dot{\mathrm{g}}_N$.

The mechanical behavior of the interface also depends on the fluid flow within it. Indeed, the total pressure p_N acting on each side $\Gamma^i_{\bar{q}}$ of the interface is defined according to the Terzaghi principle [TER 25]. It is decomposed into an effective mechanical pressure p'_N and a fluid pressure equal to the inner pressure p_{w3}:

$$\mathrm{p}_N = \mathrm{p}'_N + \mathrm{p}_{w3}. \tag{1.20}$$

In this case, all the developments applied to the mechanical contact constraint and constitutive laws in sections 1.2.1.2 and 1.2.1.4 must be treated with reference to the effective pressure p'_N rather than to the total pressure p_N. For example, equation [1.5] is recast into

$$\dot{\mathrm{p}}'_N = -\mathrm{K}_N \, \dot{\mathrm{g}}_N. \tag{1.21}$$

1.3. Numerical formulation of the element

The discretization of the governing equations is based on a segment-to-segment approach. It is suitable for large relative displacements. Fluid flows are discretized according to the triple-node approach. This method allows the modeling of a drop of pressure across the interface. This section briefly describes the formulation of the element in order to implement it in a finite element code. The exact definition of the force vectors and stiffness matrix components, as well as rotation matrix, can be found in [CER 15]. The element was implemented in the finite element code LAGAMINE, developed at the University of Liège.

1.3.1. *Space and fluid pressure discretization*

The proposed 3D coupled interface finite elements are isoparametric and quadrangular [ZIE 00]. A complete representation of the interface requires 12 nodes, as shown in Figure 1.10:

– nodes 1, 2, 3, 4: the first side of the interface $\Gamma^1_{\bar{q}}$, three mechanical (coordinates x, y, z) and a pore water pressure (p_w) degree of freedom per node;

– nodes 5, 6, 7, 8: the inner nodes of the interface Ω^3 and a pore water pressure (p_w) degree of freedom per node;

– nodes 9, 10, 11, 12: the second side of the interface $\Gamma^2_{\bar{q}}$, three mechanical (coordinates x, y, z) and a pore water pressure (p_w) degree of freedom per node.

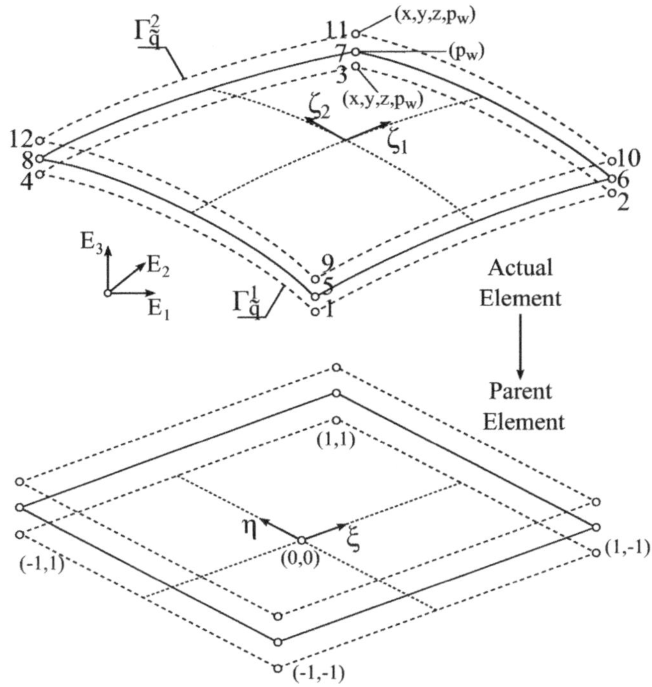

Figure 1.10. *Discretization of the interface into isoparametric elements from convective ($\zeta_1\zeta_2$) to local coordinates (ξ,η). Transformation to the parent element (source: [CER 15])*

Mechanical and hydraulic degrees of freedom are gathered into the vector of generalized coordinates at each node i such that

$$\mathbf{u}^i = \left[\mathrm{x}^\mathrm{i}, \mathrm{u}^\mathrm{i}, \mathrm{z}^\mathrm{i}, \mathrm{p}_w^i\right]^T \qquad i = 1,2,3,4,9,10,11,12 \qquad [1.22]$$

$$\mathrm{u}^i = \left[\mathrm{p}_w^i\right] \qquad i = 5,6,7,8. \qquad [1.23]$$

These coordinates are continuously interpolated over the element using typical linear interpolation functions $\phi^i(\xi,\eta)$ related to each node i of the interpolated side. Continuous generalized velocities $\dot{\mathbf{u}}$ are interpolated over the element accordingly from nodal values $\dot{\mathbf{u}}^i$.

1.3.2. *Mechanical problem*

1.3.2.1. *Local system of coordinates and gap function*

The first step of the mechanical formulation is to determine the local system of coordinates. The rotation matrix $\mathbf{R}$ relates the global $(\mathrm{E}_1, \mathrm{E}_2, \mathrm{E}_3)$ to the local

$(\mathrm{e}_1^1, \mathrm{e}_2^1, \mathrm{e}_3^1)$ system of coordinates. This rotation matrix is computed with respect to the side $\Gamma^1_{\bar{q}}$, which is the mortar side. Let us first consider the components in global axes of two unit non-orthogonal vectors respectively parallel to each edge of an element, namely

$$\mathbf{e}_\xi = \frac{1}{\sqrt{\left(\frac{\partial \mathrm{x}}{\partial \xi}\right)^2 + \left(\frac{\partial \mathrm{y}}{\partial \xi}\right)^2 + \left(\frac{\partial \mathrm{z}}{\partial \xi}\right)^2}} \left[\frac{\partial \mathrm{x}}{\partial \xi}, \frac{\partial \mathrm{y}}{\partial \xi}, \frac{\partial \mathrm{z}}{\partial \xi}\right]^T, \qquad [1.24]$$

$$\mathbf{e}_\eta = \frac{1}{\sqrt{\left(\frac{\partial \mathrm{x}}{\partial \eta}\right)^2 + \left(\frac{\partial \mathrm{y}}{\partial \eta}\right)^2 + \left(\frac{\partial \mathrm{z}}{\partial \eta}\right)^2}} \left[\frac{\partial \mathrm{x}}{\partial \eta}, \frac{\partial \mathrm{y}}{\partial \eta}, \frac{\partial \mathrm{z}}{\partial \eta}\right]^T. \qquad [1.25]$$

The normal to the element is given by the cross product:

$$\mathbf{e}_1^1 = \mathbf{e}_\xi \times \mathbf{e}_\eta. \qquad [1.26]$$

The first tangential direction $\mathbf{e}_2^1$ is identical to $\mathbf{e}_\xi$ and the second tangential direction is their cross product:

$$\mathbf{e}_3^1 = \mathbf{e}_1^1 \times \mathbf{e}_2^1. \qquad [1.27]$$

Therefore, the rotation matrix is the assembling of these vectors:

$$\mathbf{R} = \left[\mathbf{e}_1^1 \; \mathbf{e}_2^1 \; \mathbf{e}_3^1\right]. \qquad [1.28]$$

According to the continuous equation [1.1], the gap function at each point of $\Gamma^e_{\bar{q}}$ is computed according to

$$\dot{\mathrm{g}} = \begin{bmatrix} \dot{\mathrm{g}}_N \\ \dot{\mathrm{g}}_{T,1} \\ \dot{\mathrm{g}}_{T,1} \end{bmatrix} = [\mathbf{R}]^T \cdot \begin{bmatrix} \dot{\mathrm{x}}^2 - \dot{\mathrm{x}}^1 \\ \dot{\mathrm{y}}^2 - \dot{\mathrm{y}}^1 \\ \dot{\mathrm{z}}^2 - \dot{\mathrm{z}}^1 \end{bmatrix} = [\mathbf{R}]^T \cdot \Delta \dot{\mathrm{x}}. \qquad [1.29]$$

The norm of the Jacobian of the transformation of the element from the convective system of coordinates (ζ_1, ζ_2) to the isoparametric system (ξ, η) can be written as

$$\|\mathbf{J}\| = \sqrt{\left(\frac{\partial \mathrm{x}}{\partial \xi}\right)^2 + \left(\frac{\partial \mathrm{y}}{\partial \xi}\right)^2 + \left(\frac{\partial \mathrm{z}}{\partial \xi}\right)^2} \sqrt{\left(\frac{\partial \mathrm{x}}{\partial \eta}\right)^2 + \left(\frac{\partial \mathrm{y}}{\partial \eta}\right)^2 + \left(\frac{\partial \mathrm{z}}{\partial \eta}\right)^2}. \qquad [1.30]$$

1.3.2.2. *External energetically equivalent nodal forces*

The mechanical contribution of a single interface element to the external virtual power expression is derived from the continuous equation [1.12]. The energetically equivalent nodal forces associated with node *i* of the interface element are computed numerically using a Gauss scheme. For example, the mechanical nodal forces acting on the boundary of Ω^1 are computed according to

$$\mathbf{F}_E^i = \sum_{IP=1}^{n_{IP}} \left[\mathbf{R} \cdot \mathbf{t}\, \phi^i \, \|\mathbf{J}\| \, \mathrm{W}\right]_{IP}, \qquad [1.31]$$

where $\mathbf{t}$ is the force vector in local coordinates defined in equation [1.3], ϕ^i is the interpolation function associated with node i and the expression between brackets is evaluated in each of the n_{IP} integration points, associated with the Gauss weight W. The reaction forces acting on Ω^2 are computed accordingly.

1.3.3. *Flow problem*

1.3.3.1. *Internal energetically equivalent nodal forces*

Fluid flow inside the equivalent porous medium Ω^3, along the interface, involves energetically equivalent internal forces. This component is derived from equation [1.17]. It is numerically computed according to

$$\mathrm{F}_I^i = \sum_{IP=1}^{n_{IP}} \left[\left(\dot{\mathrm{S}}\, \phi^i - \mathrm{f}_{wl1}\, \nabla_{\mathbf{e}_2^1}\left(\phi^i\right) - \mathrm{f}_{wl2}\, \nabla_{\mathbf{e}_3^1}\left(\phi^i\right)\right) \|\mathbf{J}\| \, \mathrm{g}_N \, \mathrm{W}\right]_{IP}. \qquad [1.32]$$

1.3.3.2. *External energetically equivalent nodal forces*

Transversal fluid flows between Ω^1, Ω^2 and Ω^3 provide energetically equivalent external nodal forces related to fluid degrees of freedom. The contribution to the external virtual power corresponding to Ω^3 is derived from equation [1.17]. For example, it is numerically computed on the boundary of Ω^1 according to

$$\mathrm{F}_E^i = \sum_{IP=1}^{n_{IP}} \left[\left(\rho_w \, \mathrm{T}_{w1} \left(\mathrm{p}_{w1} - \mathrm{p}_{w3}\right) \phi^i - \rho_w \, \mathrm{T}_{w2} \left(\mathrm{p}_{w3} - \mathrm{p}_{w2}\right) \phi^i\right) \|\mathbf{J}\| \, \mathrm{W}\right]_{IP}, \qquad [1.33]$$

where p_{w1} is the fluid pressure on side 1, p_{w2} on side 2 and p_{w3} inside. The reaction forces acting on the boundary of Ω^2 are computed similarly.

1.3.4. *Time discretization*

Internal $\mathbf{F}_I^i$ and external $\mathbf{F}_E^i$ nodal forces defined in equations [1.31], [1.32] and [1.33] are gathered into the global vectors $\mathbf{F}_I$ and $\mathbf{F}_E$. Therefore, the vector of out-of-balance forces $\mathbf{F}_{OB}$ is defined according to

$$\mathbf{F}_{OB} = \mathbf{F}_I - \mathbf{F}_E. \qquad [1.34]$$

The fluid flow problem within a porous medium is inherently time dependent. Therefore, modeling its evolution requires the discretization of time. It is assumed that the media in contact are initially in equilibrium at a given time t, i.e. $\mathbf{F}_{OB} = \mathbf{0}$. The equilibrium of the discretized system should be verified over a whole time step Δt such that

$$\int_t^{t+\Delta t} \mathrm{G}(t)\, \mathbf{F}_{OB}\, \mathrm{d}t = \mathbf{0} \qquad [1.35]$$

where $\mathrm{G}(t)$ is a weighting function. In this work, the weighting function is reduced to a collocation $\delta(\theta)$, where δ is the Dirac function. It is proved that a choice of $\theta \geq 0.5$ leads to an unconditionally stable time scheme [ZIE 00]. In this work, the integration scheme is implicit, i.e. $\theta = 1$. The equilibrium is then written at the end of the time step.

1.3.5. *Stiffness matrix*

The stiffness matrix $\mathbf{K}$ related to the interface element is computed analytically by deriving the out-of-balance forces related to node i with respect to the generalized degree of freedom j. The extended developments are provided in [CER 15].

1.4. Application

The soil–foundation interaction plays a major role in the resistance of different types of offshore foundations (e.g. piles, anchors, suction caissons) to combined horizontal, moment and tension loads. Suction caissons are an emerging technology for the foundation of offshore wind turbines. Their transient behavior is inherently linked to hydro-mechanically coupled phenomena. Therefore, they represent a suitable example of interface element capabilities.

1.4.1. *Suction caissons*

Ambitious objectives of the European Union to increase the share of renewable energy in the Union's energy mix have led to the multiplication of offshore wind energy projects. The design of a suitable foundation is challenging and may represent one third of the total project's cost [BYR 03, SEN 08]. While monopiles represent the great majority of installed foundations in shallow waters, other types of foundations such as tripod or jacket structures are used in deeper waters and can be coupled with different types of foundations.

Among different types of foundations, suction caissons, also termed bucket foundations or suction anchors, should be highlighted [TJE 90, ISK 02, HOU 05c,

LI 15]. They consist of a hollow cylinder that is open towards the bottom. Their top (the lid) can be a stiffer plate or a dome [TRA 05]. They are mainly made of steel and may have stiffeners added inside to reinforce the structure. They are used as temporary anchors [SEN 82] or permanent foundations [TJE 90] in sandy [IBS 14] or clayey soils [GOU 09].

Installation of suction caissons requires only light equipment [HOU 05a]. Initially, the caisson penetrates the sea bed under its own weight. Afterwards, the water trapped inside is pumped out through an opening. A differential of water pressure between the inside and the outside of the caisson digs it into the soil, as shown in Figure 1.11. In addition, the seepage flow created reduces penetration resistance at the tip and along the inner skirt, facilitating the installation [AND 08, SEN 09, BIE 18b, RAG 19]. The foundation may be reused by reversing the process.

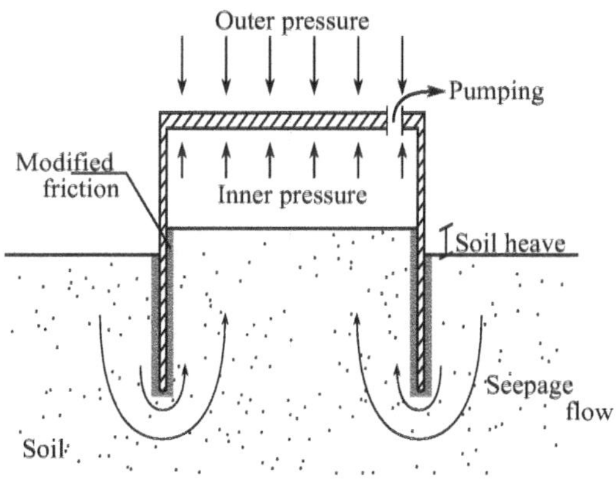

Figure 1.11. *Sketch of the installation process*

Offshore wind structures are very light but subject to a large overturning moment. Therefore, attention is paid to the definition of failure criteria under combined horizontal load and moment in different configurations [BYR 02, GOU 08, BRA 09, IBS 14, LI 15]. When the superstructure is a multipod, the large moment is converted into a push-pull vertical load acting on the caissons. In this case, the ability to resist large tension load is fundamental [HOU 05b, HOU 06, GAO 13, BIE 18b] and the suction effect may be transiently mobilized for this purpose.

The behavior of suction caissons results from the interaction between the caisson and the seabed. The soil behavior is the first source of complexity and nonlinearity [THI 14]. Moreover, the mechanical problem is coupled with a complicated fluid flow around the caisson. Depending on the geometry, soil permeability and loading rate, the behavior of the soil is never completely drained or undrained, but most of the time, partially drained [THI 14, CER 16]. However, in all cases, the strength of the caisson depends strongly on the soil–caisson interface behaviour and especially its coefficient of friction [ACH 13, KOU 14, CER 16].

Modeling of a suction caisson is, therefore, a perfect example to illustrate the basic working of hydro-mechanically coupled interface elements. The following simulations involve mechanical contact/loss of contact, friction mobilization, sliding, transversal/longitudinal fluid flow and storage of water inside a gap. Since this chapter focuses on the interface behavior, the loading conditions, constitutive law of the soil (simple elastoplastic) and flow modeling (fully saturated hypothesis, constant permeability) are kept as basic as possible. The behavior of the caisson is explained by vertical and lateral imposed displacements at the top of the caisson. The role of interface elements is highlighted in each simulation.

1.4.2. *Problem description*

A steel suction caisson is assumed to be embedded in a homogeneous sand layer, as shown in Figure 1.12a. Vertical and lateral loadings are investigated in the following using the same 3D mesh for consistency, although axisymmetric conditions are sufficient to study the vertical loading. The soil is represented by half of a cylindrical layer. Its radius (Rsoil) is equal to 24 m and its height (Hsoil) is equal to 12 m, as shown in Figure 1.12a. The finite element mesh is composed of 16,058 nodes and 14,190 elements, including volume (eight nodes, linear interpolation functions), interface (four nodes) and boundary elements. Geometrical parameters of the mesh are given in Table 1.2.

The caisson is composed of a horizontal lid at the top and a vertical skirt, as shown in Figure 1.12b. The ratio of the skirt thickness (t_{skirt}) to the diameter ($2\,R_{out}$) is larger than the real one [BYR 02, KEL 06] to avoid numerical issues. Indeed, using thin and elongated elements may lead to numerical instabilities, especially during the iterative solving of the global matrix. The effect on the results is limited since the bearing capacity of the tip of the caisson is not mobilized during tension simulations and only partly mobilized during lateral loading.

The mudline delineates the solid and liquid phases at the bottom of the sea. The reference level of the sea top is chosen as 10 m above the mudline. However, this does not matter in this simulation. What matters is only the variations of pore water pressures (PWP) to balance tension and lateral loading. Above all, no cavitation is taken into account in the flow modeling. The lateral face (y-z plane) is considered impervious because of the symmetry. The lower face (x-y plane) is also deemed impervious and could represent the transition to another layer of much lower permeability. Pore water pressures are fixed on the other faces and are equal to the hydrostatic pressure.

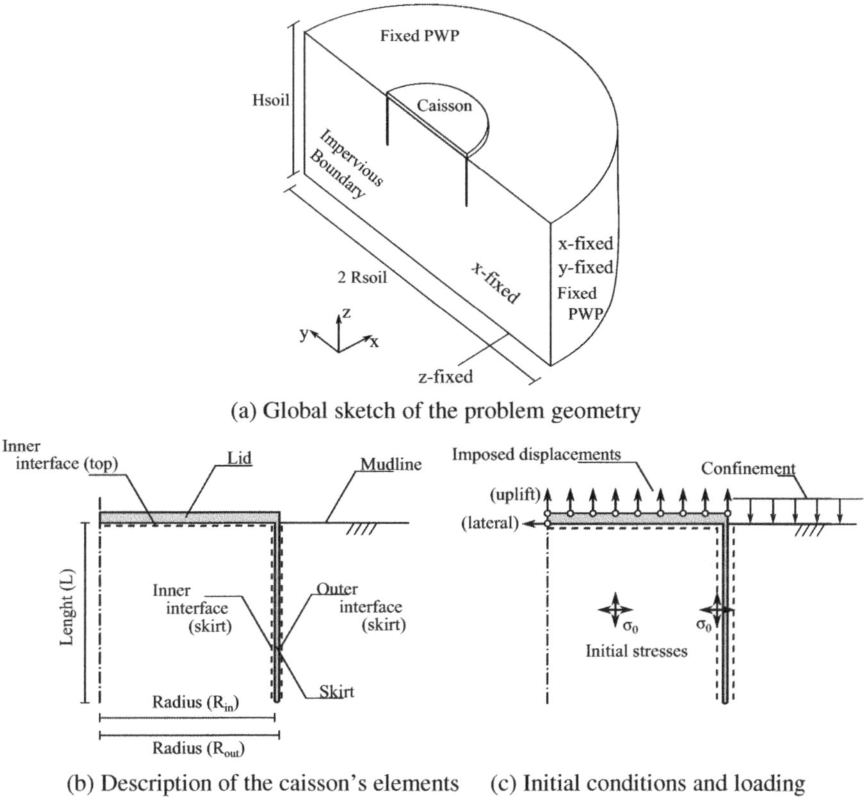

Figure 1.12. *Statement of the problem*

Caisson					Soil		Mesh	
R_{int}	R_{out}	L	t_{skirt}	t_{lid}	R_{soil}	H_{soil}	N_{nodes}	N_{elems}
3.8 m	3.9 m	4 m	0.1 m	0.4 m	24 m	12 m	16,058	14,190

Table 1.2. *Geometrical parameters:* R_{int}, *inner radius;* R_{out}, *outer radius; L, length;* t_{skirt}, *thickness of the skirt;* t_{lid}, *thickness of the lid;* R_{soil}, *outer radius of the soil domain;* H_{soil}, *thickness of the soil layer;* N_{nodes}, *number of nodes;* N_{elems} *number of elements*

The installation of the caisson into the soil is not modeled. This could have large implications on the behavior of the caisson, since the stresses are modified and the soil is disturbed by the installation [RAG 19]. However, for the sake of simplicity, initial effective stresses (in the soil and interface) corresponds to the weight of the soil (coefficient of earth pressure at rest K_0 is equal to 1). The hydrostatic pore water pressures corresponding to the depth of water are initialized.

The mechanical behavior of the caisson is always elastic, but two distinct hypotheses are considered for the soil: elastic and elastoplastic. The elastic law considers a relatively high Young modulus of 200 MPa. The elastoplastic law is an internal friction model, PLASOL, implemented in the finite element code LAGAMINE [BAR 98]. The Young modulus is equal to the elastic one and characterizes the low deformation behavior. Its initial friction angle ϕ_i is equal to 5° and may harden up to $\phi_f = 30°$. The hardening rule is characterized by the parameter B_ϕ, indicating that half of the hardening is reached for a deformation equal to B_ϕ. The porosity n and the specific mass γ_s are identical for the soil and the steel in order to ensure a problem initially in equilibrium. The permeability is equal to 1.E-4 m/s. All material parameters are summarized in Table 1.3.

The transversal conductivity T_w that characterizes the interface is null between the caisson and the interface (the caisson is impervious) but not null between the soil and the interface. The residual hydraulic aperture D_0 is equal to 5.E-5 m. Therefore, there is always a longitudinal fluid flow event in case of contact. The soil caisson friction coefficient is equal to 0.50. The penalty coefficient is fixed at 1.E10 N/m^3, and the influence of this parameter is investigated in the following.

Soil							
E [MPa]	ν [-]	n [-]	k [m/s]	γ_s [kg/m^3]	ϕ_i [°]	ϕ_f [°]	B_ϕ [-]
2E2	0.3	0.36	1.E-4	2,650	5	30	0.005
Caisson							
E [MPa]	ν [-]	n [-]	k [m/s]	γ_s [kg/m^3]			
2E5	0.3	0.36	0	2,650			
Interface							
K_N [N/m^3]	K_T [N/m^3]	T_w [m.Pa^{-1}.s^{-1}]		D_0 [m]	μ [-]		
1E10	1E10	1.E-8		1.E-5	0.50		

Table 1.3. *Material parameters: E, Young's modulus; ν, Poisson's ratio; n, porosity; k, permeability; γ_s, density of solid grains; K_0, coefficient of earth pressure at rest; K_N, K_T, penalty coefficients; μ, friction coefficient; T_w, transversal conductivity; D_0, residual hydraulic aperture*

First, the simulations investigate the uplifting behavior of the suction caisson. A uniform vertical displacement is imposed over the whole upper surface of the caisson. The tension load is balanced only by the shear along the shaft of the caisson and suction effect. The total load ΔF_{tot} applied[1] to the caisson is the sum of all the reactions computed at each imposed degree of freedom. ΔF_{in} and ΔF_{out} are respectively the integral of shear stresses acting on the inner and outer walls of the

1 Results from the simulations are multiplied by 2 to correspond to an entire caisson.

caisson, as shown in Figure 1.13. Finally, the variation of PWP under the top of the caisson is integrated and summarized into ΔF_{pw}.

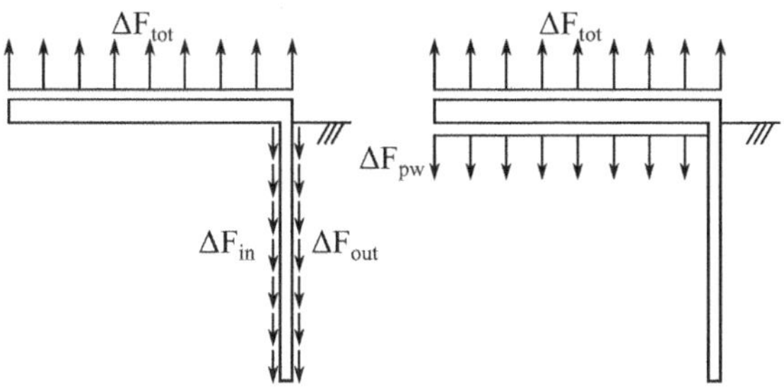

Figure 1.13. *Components of reaction balancing a tension load: shearing inside and outside the shaft (left) and pore water pressure variation (right)*

Second, a lateral displacement is imposed at the center of the lid of the caisson. In this case, the movement of the caisson is not axisymmetric at all. The caisson rotates, shear stresses act inside and outside the caisson and passive earth pressures are mobilized in front of the caisson in the direction of loading, as shown in Figure 1.14. In this case, ΔF_{tot} is the lateral reaction at the center, but the other components are kept identical.

The distribution of shear stresses, effective/pore water pressures and p-y curves will be defined to better understand the behavior of the caisson. These p-y curves result from the idealization of the laterally loaded pile–soil interaction problem as a beam supported by nonlinear springs, representing the soil reaction [MCC 56, REE 10]. They relate the resultant total horizontal load per meter length of the pile, applied by the soil on the caisson, as a function of its lateral displacement y.

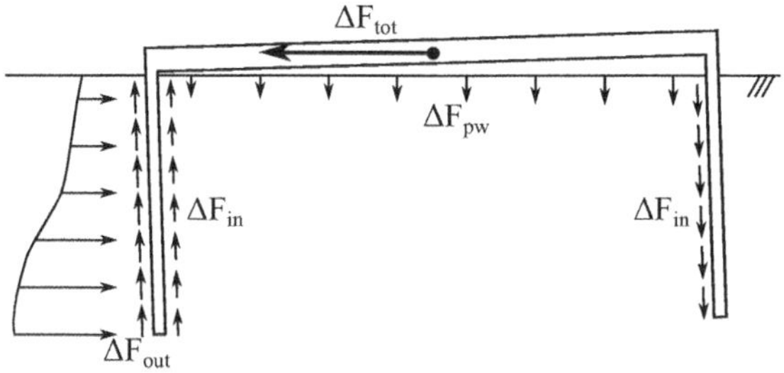

Figure 1.14. *Components of reaction balancing a lateral load: shearing inside and outside the shaft and pore water pressure variation*

1.4.3. *Tension loading*

1.4.3.1. *Drained simulation (mechanical problem)*

During a drained uplift simulation (elastic soil), the total tension load applied is balanced only by shearing along the shaft and the weight of the caisson. Figure 1.15 presents the evolution of shear components with the applied vertical displacement. Variations of inside and outside shearing are almost linear at the beginning, since variations of shear stress are given by

$$\dot{\tau} = \mathrm{K}_T \dot{\mathrm{g}}_T \qquad if \qquad \tau < \mu\, \mathrm{p}'_N. \qquad [1.36]$$

At the very beginning, this criterion is not met anywhere and variation of shear stress is almost identical all along the shaft. The interface is still in a *stick state*.

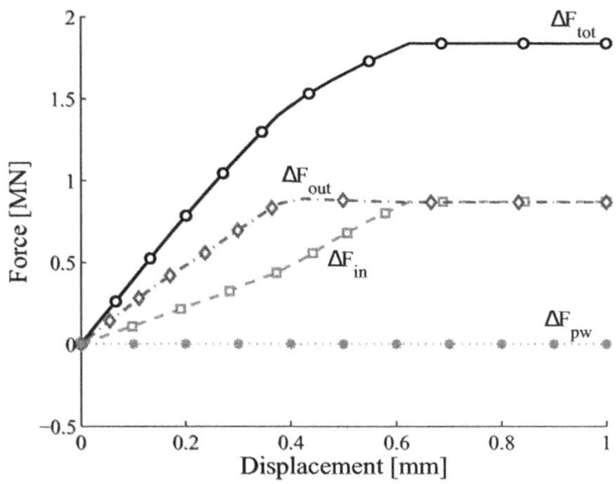

Figure 1.15. *Drained uplift simulation of the caisson: components of reaction for the entire caisson*

The force–displacement result is stiffer for the outside shear component ΔF_{out} than the inside shear component ΔF_{in}. Subsequently, the maximum shear stress is mobilized first all along the outside wall of the shaft after an imposed displacement of 0.4 mm. Any increment of the tension load applied is balanced by shearing along the inside wall, inducing a slight slope breakage in the ΔF_{tot} and ΔF_{in} results. The maximum total tension is reached after 0.6 mm, when the maximum shear stress is reached all along the inside wall. There are no variations of pore water pressures since the simulation is drained ($\Delta F_{pw} = 0$).

The stiffness of shear mobilization is different inside and outside the caisson. The soil inside the caisson is more confined than that outside, and tends to move along vertically with the caisson, as shown in Figure 1.16. Subsequently, the relative tangential displacement rate $\dot{g}_t$ along the inner interface is slower than the outside, which affects the shear rate $\dot{\tau}$. Finally, the initial slope of ΔF_{in} is lower than ΔF_{out}.

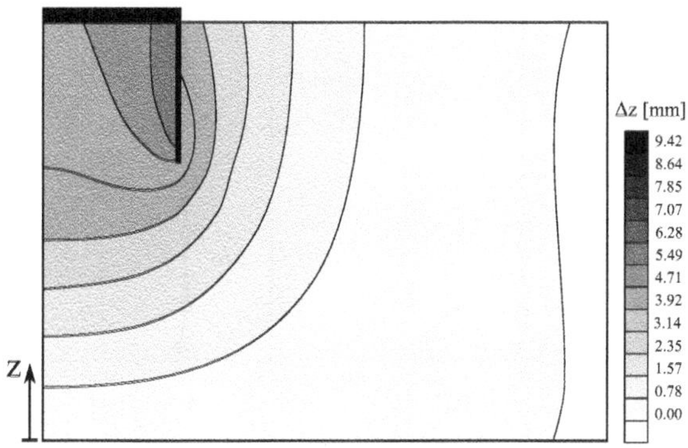

Figure 1.16. *Drained uplift simulation of the caisson: cross-section of the vertical displacement, imposed vertical displacement at the top of the caisson* = 1 *mm, drawing in the initial configuration*

The global behavior of the caisson is closely linked to the shear stress distribution along its shaft. A vertical cross-section along the 4 m length shaft inside the caisson is shown in Figure 1.17 for different imposed vertical displacements. The vertical shear stress (τ_z) is shown on the left and the reduced shear stress ($\tau_z/p'_N \leq \mu$) on the right. After the full mobilization of friction ($\tau_z/p'_N = \mu = 0.5$), the distribution of shear stress increases with depth, due to an increasing normal pressure p'_N. Hence, the maximum shear stress is first mobilized at the top of the caisson (low maximum stress) and, finally, at its bottom.

In this figure, τ_z is equal to zero at the top of the caisson after a sufficient displacement. This indicates the creation of a gap between the caisson and the soil. In this case, p'_N is equal to zero and shearing is not possible. This gap is due to the stress distribution within the soil, resulting from shear stress mobilization along the interface, which tends to separate the soil from the caisson. The length of the gap depends on physical parameters (relative stiffness of the soil and caisson) and non-physical parameters (penalty coefficient, which is detailed in the following). Outside the caisson, the gap is longer (1.5 m) and occurs faster since the soil is not confined and is free to move horizontally. The size of this gap is a function of the soil parameters, as a softer soil would be more prone to collapse and avoid gaping.

The stiffness difference between inside and outside friction mobilization does not affect the maximum load that can be sustained by the caisson. However, it affects the small displacement behavior and then serviceability.

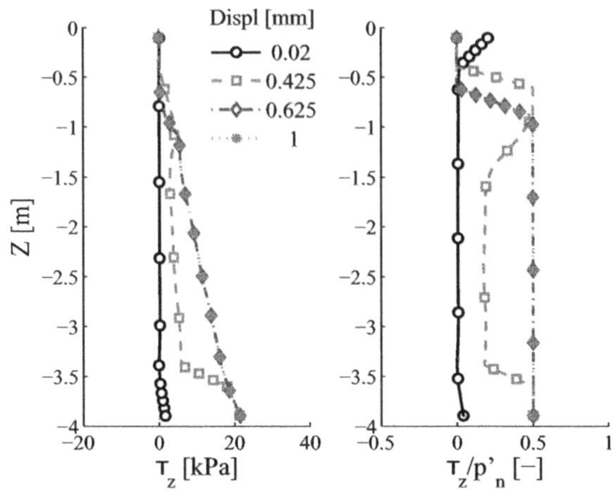

Figure 1.17. *Drained uplift simulation of the caisson: cross-section along the shaft for different vertical displacements (inside)*

As discussed previously, two penalty coefficients (K_N and K_T) are introduced in the formulation of the interface finite element. Therefore, it is important to assess their influence on the results. Four orders of magnitude of normal penalty coefficients are compared in Figure 1.18a. The results prove that the behavior is only slightly affected by these variations. At low displacement, the higher the K_N value, the higher the stiffness. For coefficients lying between 1.E09 and 1.E11, the final tension loads increase slightly. However, the lowest coefficient leads to a higher final value.

For a drained simulation, the penalty coefficient mainly affects the gap opening behavior since the normal pressure rate is given by $\dot{p}'_N = K_N \, \dot{g}_N$. A gap is detected when this pressure is equal to zero. Therefore, reducing the penalty coefficient leads to less extended gaps and larger zones for shear stress mobilization. However, the normal pressure distribution is also modified along the shaft due to the global movement of the soil. Therefore, the final results are almost not affected in this case since these effects compensate for each other.

It must be noted that soil behavior may have a much greater impact on the results than penalty coefficients. In purely drained conditions, the maximum tension load is identical but stiffness is affected by varying the material parameters. In the reference case, the maximum load is reached after 0.6 mm. It increases up to 6 mm if an elastic soil (E =20 MPa) is considered instead. However, it is limited to 0.9 mm if the elastoplastic constitutive law is used, suggesting that plastic zones are very limited in this case. Therefore, the error introduced by the penalty method is lower than approximations on the real soil behavior.

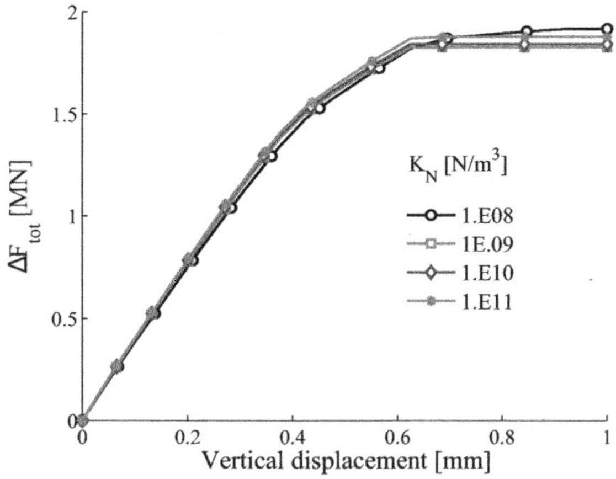

(a) Influence of the normal penalty coefficient

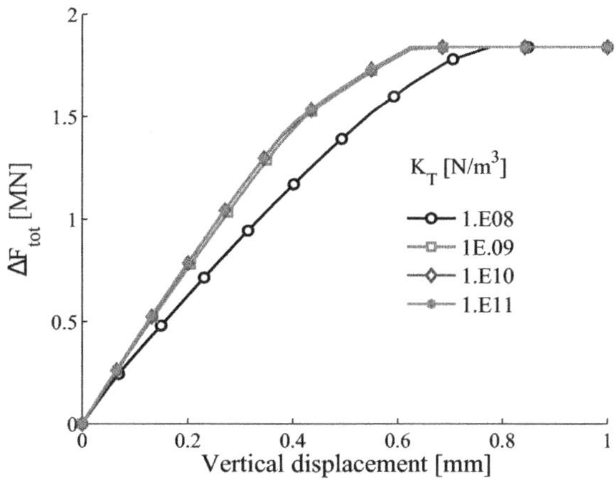

(b) Influence of the tangential penalty coefficient

Figure 1.18. *Drained uplift simulation of the caisson*

The influence of the K_T coefficient on the results is assessed in Figure 1.18b. In this case, the final value is not altered but the stiffness is. Contrary to the K_N coefficient, which is completely non-physical and should be as high as possible, K_T may have a physical signification. In other words, as K_T decreases, the displacement that should be applied to reach the maximum shear stress increases. The actual interface behavior is more complicated, as it is nonlinear, normal stress dependent and includes dilatancy. Subsequently, the accurate simulation of this behavior would require dedicated complex constitutive laws, such as [STU 16].

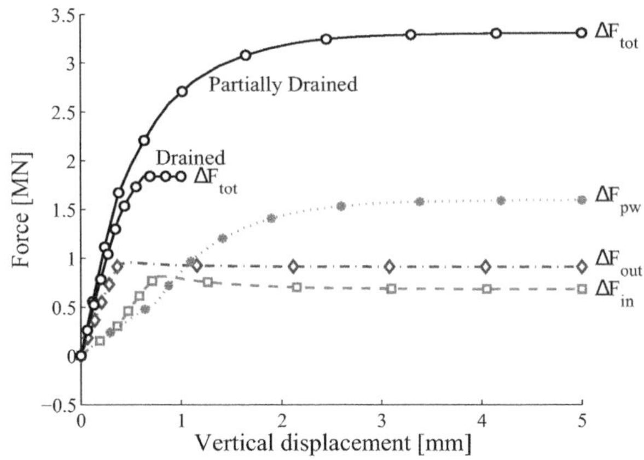

(a) Loading rate 0.05 mm/s, drained total force is added for comparison

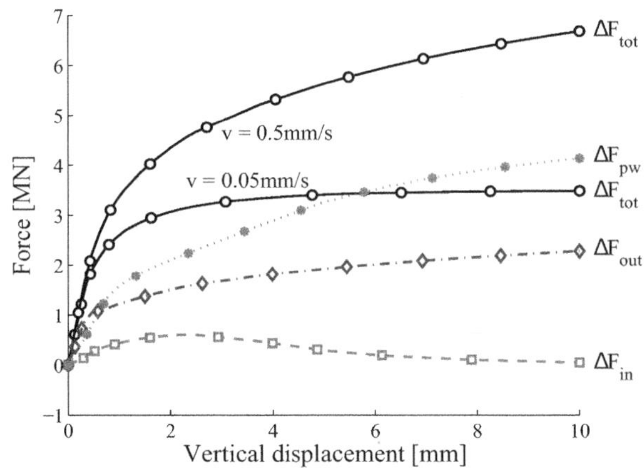

(b) Loading rate 0.5 mm/s, total load for a rate of 0.05mm/s is added for comparison

Figure 1.19. *Partially drained uplift simulation of the caisson: components of reaction for the entire caisson*

As a partial conclusion, it was shown that the drained tension behavior of a caisson is strongly dependent on interface properties, much more than the soil volume behavior. Friction mobilization is quite important and could only be represented by adequate finite elements. The following section introduces hydro-mechanical couplings and explains why the proposed coupled element is necessary.

1.4.3.2. *Partially drained simulation (coupled problem)*

The results of a partially drained simulation (elastic soil) are presented in Figure 1.19a. For comparison, the drained and partially drained total loads ΔF_{tot}

applied are shown together. The two partially drained shear components are presented as well as the pore water pressure variation component ΔF_{pw}. The total capacity is almost doubled, ductility is increased and the response is stiffer than the drained simulation. The capacity is reached after 0.6 mm in drained conditions and 3 mm in partially drained conditions.

Detailed results at a faster loading rate (0.5 mm/s) are presented in Figure 1.19b and compared to the total load evolution at a slower uplifting rate (0.05 mm/s). Compared to the simulation presented in Figure 1.19a, the suction component ΔF_{pw} is larger as well as the displacement at the steady state, as PWP have less time to dissipate. In addition, the inside shear component tends to zero at large displacement, indicating that the soil is plugged inside the caisson (typical of undrained behavior). There is then no relative displacement and no shearing. Finally, outside shearing is increased due to the negative water pressure that tends to keep soil and caisson edges close enough.

The caisson behavior changes from completely drained to undrained, as the uplifting rate increases (or the permeability decreases). This rate effect is well known and was modeled previously by Thieken *et al.* [THI 14], although the gap formation is simulated using "water elements". The more undrained the loading, the higher the (transient) capacity and stiffness. However, PWP variations become very negative, and it must be verified whether some hypotheses are still valid, such as the full saturation of the gap and the absence of water cavitation.

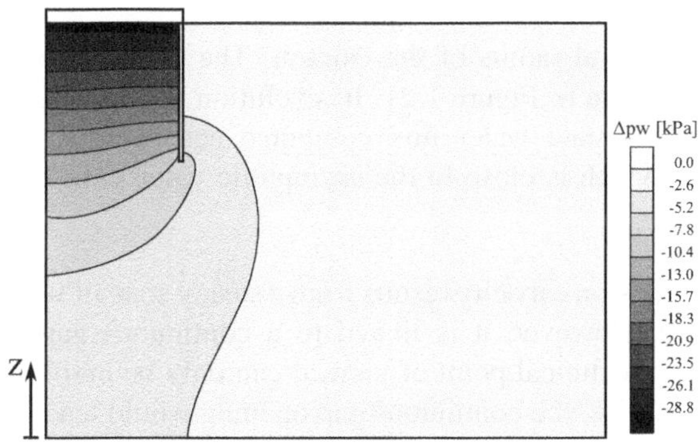

Figure 1.20. *Partially drained uplift (loading rate 0.05 mm/s) simulation of the caisson: cross-section of pore water pressure variation, imposed vertical displacement at the top of the caisson* $= 5$ *mm, drawing in the initial configuration*

Variations of PWP within the soil are responsible for these different behaviors. The uplifting of the caisson creates a gap under the lid. It is assumed that this gap is fully saturated with water, inducing a fluid flow from the soil to the gap and resulting in the negative variations of PWP, as shown in Figure 1.20 (loading rate is equal to 0.05 mm/s). The most negative variation of pressure is just under the lid of the caisson. Therefore, the total water pressure over the lid is greater than that under the lid. This difference of pressure creates the suction effect, transiently balancing a part of the tension load.

This suction component $\Delta \mathrm{F}_{pw}$ is the main contributor to the strength improvement, as shown in Figure 1.19a. However, it is not mobilized simultaneously to the friction components. At low displacement, shearing along the shaft is the larger contributor to the tension strength. $\Delta \mathrm{F}_{pw}$ starts increasing significantly when friction is mobilized inside and outside. It finally reaches a plateau. This evolution is correlated with the gap opening under the caisson's lid. The gap opening rate v_g depends on

$$v_g = \dot{\Delta} z_c - \dot{\Delta} z_s, \qquad [1.37]$$

where $\dot{\Delta} z_c$ and $\dot{\Delta} z_s$ are the velocity of material points of the caisson and the soil, at the interface under the lid. The gap opening induces negative variations of PWP and an inverse consolidation process. The soil is uplifting $\dot{\Delta} z_s > 0$ but slower than the caisson. The water flow, soil and caisson movements interact together, but finally reach a steady state where $\dot{\Delta} z_s \to 0$. Finally, the gap opening and the caisson velocity are equal. Therefore, the water flux towards the gap is computed (if the caisson is deemed completely rigid) according to

$$f_g = \rho_w \, v_g \, \pi \, \mathrm{R}_{in}^2. \qquad [1.38]$$

where R_{in} is the internal radius of the caisson. The evolution of the total flux f_g towards the gap is shown in Figure 1.21. Its evolution is homothetic to the evolution of $\Delta \mathrm{F}_{pw}$. The steady-state water flux computed according to equation [1.38] is equal to 2.27 kg/s, which is close to the asymptotic value computed numerically in Figure 1.21.

The increase in tension capacity results from a steady state of water flow (stabilized variations of PWP). However, it is linked to a continuous gap opening and then uplifting. From a geotechnical point of view, it can only be mobilized to balance the transient load. Otherwise, the continuous gap opening would lead to the soil collapse below the moving tip and to a continuous reduction of the drainage length, eventually leading to the failure of the caisson. In addition, the longitudinal gap opening, if it is too large, may modify the results by increasing the dissipation of underpressures. All of these effects are dependent on the caisson uplifting rate and soil permeability.

The constitutive law of the soil does not affect the load–displacement relationship in all cases, as shown in Figure 1.22. The initial stiffness is similar for elastic and elastoplastic constitutive laws. However, the capacity diverges when the flow regimes approach undrained conditions. Indeed, the drained and low uplift rate behaviors are dominated by failure along the interface. On the contrary, undrained or almost undrained conditions lead to soil plugging. Failure and plasticity occur within the soil. Consequently, the constitutive law that describes the soil behavior has a great influence on the results.

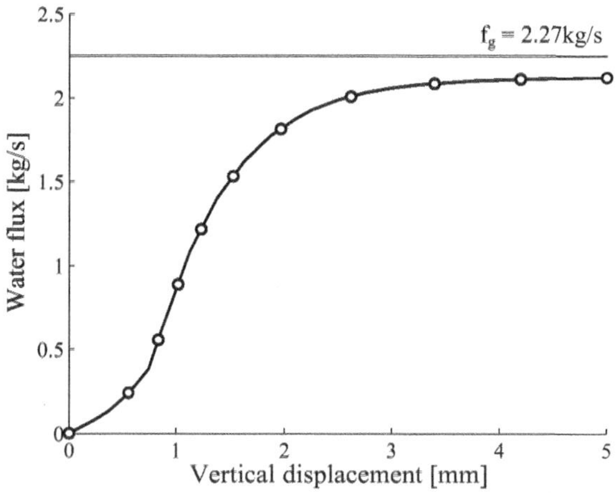

Figure 1.21. *Partially drained uplift simulation (0.05 mm/s) of the caisson: total water flux from the soil to the gap under the top of the caisson*

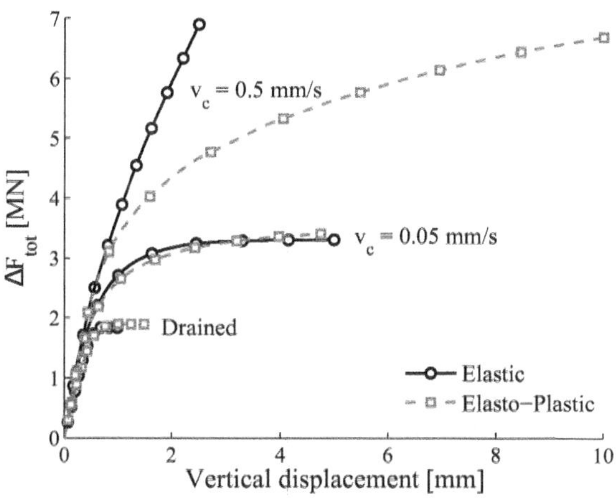

Figure 1.22. *Partially drained uplift simulation of the caisson: comparison of elastic and elastoplastic constitutive laws at different rates*

1.4.3.3. *Comparison with other authors*

In order to validate the interface element and the proposed methodology, the results presented by Thieken *et al.* [THI 14] are reproduced in Figure 1.23. Geometry (L =10 m, R_{out} = 5 m) and soil parameters (internal friction model) are different, and the details can be found in [THI 14]. The results presented for the hydro-mechanically coupled interface are very consistent with those obtained by the "water element" method.

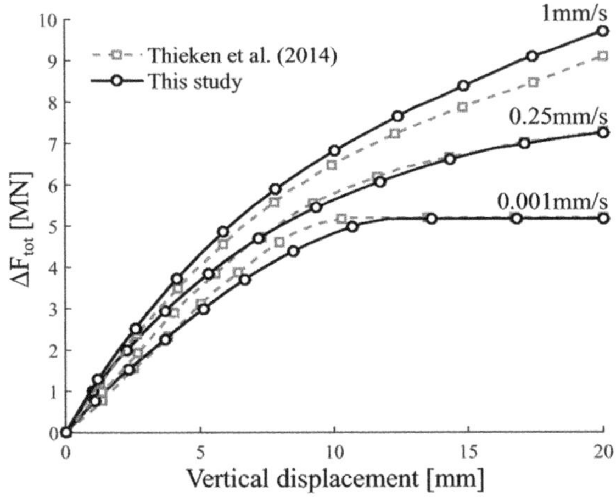

Figure 1.23. *Partially drained simulations for different permeabilities: comparison of this work and the published results from [THI 14]*

1.4.3.4. *Loading/unloading cycle*

Typical loading applied to offshore foundation is neither monotonic nor displacement controlled. It consists of a cyclic force that results from the wind and wave acting on the wind turbine and applied onto the foundation [BIE 18a, BIE 18b]. Modeling the long-term cyclic behavior of the suction caisson is out of the scope of this study and was partly addressed in [CER 16]. In the following, a single cycle loading is applied to the caisson and the different components of reaction are detailed to inform the global behavior of the caisson. The loading is inspired from the centrifuge tests presented in [BIE 18a, BIE 18b] and a cyclic amplitude of 80 kPa is applied on the caisson for a zero average load, over a period of 11 s representative of the offshore environment.

For the reference case, the results are presented in Figure 1.24. The behavior of the caisson is almost undrained, as most of the total load applied is balanced by the variations of PWP, and the internal shear stress mobilized, ΔF_{in}, is close to zero. It is interesting to note that the load–displacement relationship is not completely symmetrical. A net tensile displacement can be observed in Figure 1.24a after a single cycle, the stiffness is not identical in compression or tension and the total shear force along the interface does not come back to zero inside and outside the caisson after a cycle. However, there is almost no accumulated pore water pressure. This is consistent with observations reported in [BIE 18a, BIE 18b].

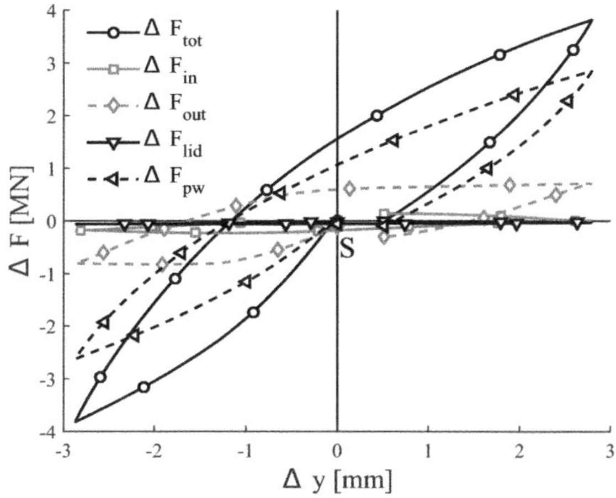

(a) Force versus vertical displacement, S = starting point

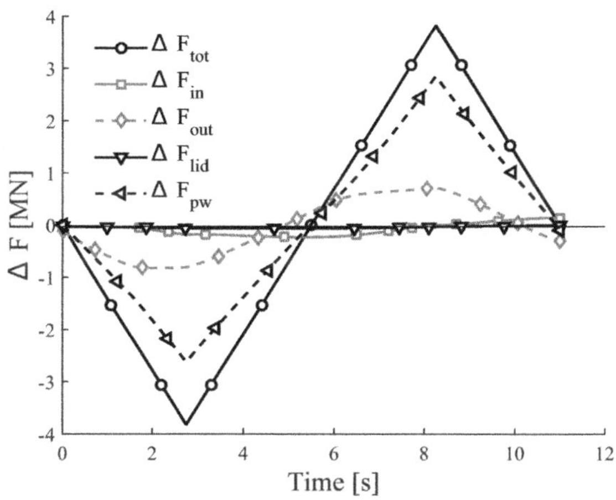

(b) Force versus time

Figure 1.24. *Simulation of a single compression/tension cycle (k = 1.E-4 m/s)*

A second simulation was run, considering a lower permeability (k = 1.E-3 m/s), to highlight how the drainage regime can affect the overall behavior of the caisson

(Figure 1.25). In this case, the PWP is still the main contribution to the caisson strength, but there is a clear contribution of the shear inside the caisson, as well as the lid (in compression). The lid contribution participates in the non-symmetry of the load–displacement relationship. Surprisingly, the contribution of PWP at the end of the loading is not null in this case and indicates a net increase.

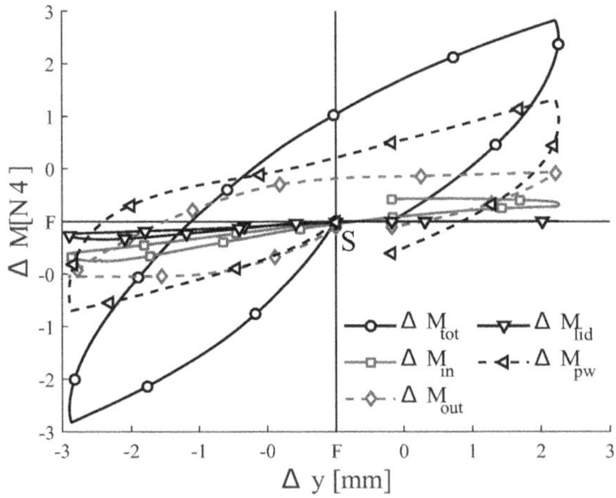

(a) Force versus vertical displacement, S = starting point

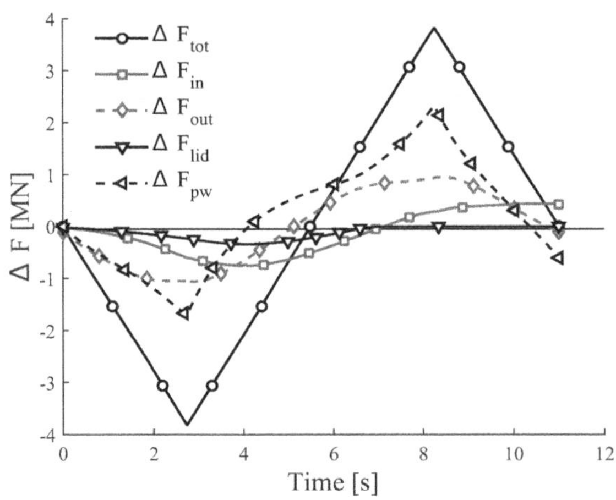

(b) Force versus time

Figure 1.25. *Simulation of a single compression/tension cycle (k = 1.E-3 m/s)*

In both cases, it was shown that the cyclic loading applied is mainly balanced by the variations of PWP, which are reversible. The effective stress variations within the soil, which generate plastic deformation, are limited. The final displacement at the end of the loading is respectively equal to 0.39 mm (k = 1.E-4 m/s) and 0.67 mm (k = 1.E-3 m/s). Consequently, the hydro-mechanical couplings can actually decrease the cyclic-induced settlement of the soil around the caisson, if the caisson is designed for that purpose, as highlighted in [CER 16].

In both cases, the observed non-symmetrical behavior is mainly due to hydro-mechanical couplings. However, the asymmetry of the results from [BIE 18a, BIE 18b] is much stronger than the results presented here. It is expected that a more complex constitutive law including Lode angle dependency would be more appropriate, as the stiffness and strength of soils are different in compression and extension.

1.4.4. *Lateral loading*

In the following, a lateral displacement is imposed at the center of a caisson in the y-direction while the displacement is free along the vertical direction. This loading mode corresponds more to anchors than to foundations (usually higher L/D ratio). However, an identical geometry is used, to ensure the consistency of this chapter. It is also shown in the following that the nonlinear behavior of the caisson depends mostly on the plasticity that develops within the soil. Therefore, the soil is only described by an elastoplastic constitutive law.

1.4.4.1. *Drained simulation (mechanical problem)*

Upon an imposed lateral displacement, the drained behavior of the caisson becomes non-symmetrical. Cross-sections of lateral and vertical soil displacements after an imposed central displacement of 2.5 mm are presented in Figures 1.26 and 1.27 respectively. The displacement of the caisson is the combination of three rigid body motions: a lateral displacement, a small vertical displacement and a rotation.

The soil in front of the caisson is pushed away along with the lateral displacement of the caisson, creating a plastic wedge. This is suggested in Figure 1.26 by the steep gradient of displacement, in front of the caisson (on the left). Plastic strains are larger near the surface of the soil, where the average stress is low, which in turn induce larger lateral displacements and eventually a rotational movement, as shown in Figure 1.27. Consequently, the shear stress developed along the caisson's shaft has two components. The horizontal shear stress is due to the lateral relative displacement, while the vertical component is due to the rotation of the caisson. The soil and caisson displacements (both lateral and vertical) are not continuous between the inside and the outside of the caisson. This suggests that a gap is created behind the caisson. No shear resistance could then be mobilized along this zone. Inside the caisson, the soil accompanies more or less the movement of the caisson, except along the top. The difference in vertical displacement along the top surface indicates that a gap is also created there.

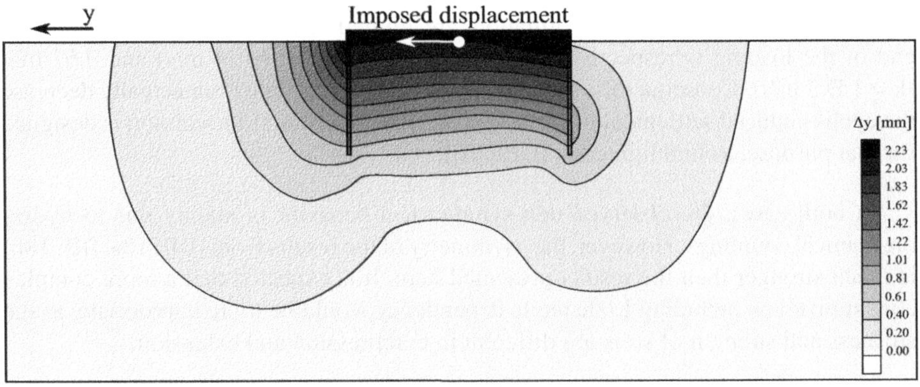

Figure 1.26. *Drained lateral displacement of the caisson after an imposed lateral displacement of 2.5 mm: lateral displacement (Δy), cross-section in the Y-Z plane, drawing in the initial configuration*

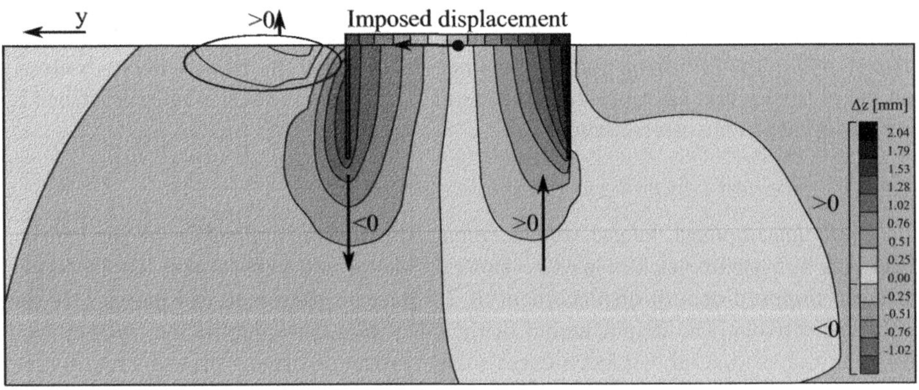

Figure 1.27. *Drained lateral displacement of the caisson after an imposed lateral displacement of 2.5 mm: vertical displacement (Δz), cross-section in the Y-Z plane, drawing in the initial configuration*

The total horizontal load applied to the caisson (ΔF_{tot}) is balanced directly by contact pressure and shear along the shaft of the caisson in the horizontal y-direction (parallel to the lateral face of the mesh). However, due to the rotational movement, non-negligible vertical shear stresses (inside ΔF_{in} and outside ΔF_{out}) are also generated within the soil–caisson interface. They are represented similarly to tension simulations in Figure 1.28. In addition, shear is mobilized at the base of the caisson. It may lead to a global failure if the shear capacity is reached, which is not observed here, as the caisson has a large diameter.

The total load–displacement relationship presents a classical shape with a clear change of slope after an imposed displacement of approximately 2 mm [ACH 10, BIE 12]. This is mainly due to the development of plasticity within the soil. The shear components have different magnitudes and signs. The direction of shear forces is upwards at the front of the caisson and downwards behind it, due to the rotation. However, the magnitude of shear stress is different in absolute value between the inside and the outside of the caisson, due to the gap opening (outside at the back). Maximum friction is reached in both cases since a plateau is reached.

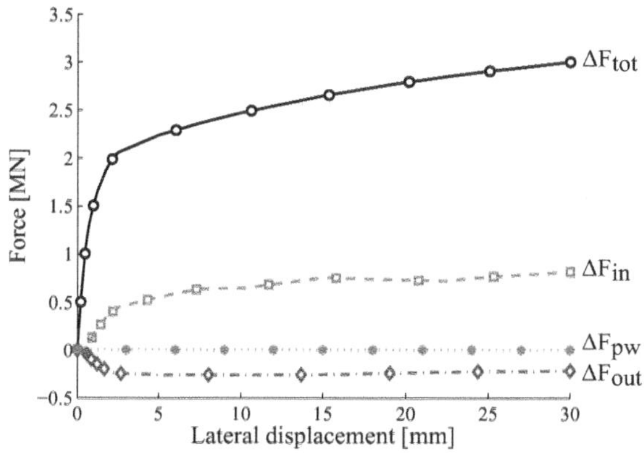

Figure 1.28. *Drained lateral simulation of the caisson: components of reaction for the entire caisson*

Cross-sections of shear and normal pressure distribution along the shaft of the caisson are presented radially in Figure 1.29. As these values are vectors, they are therefore characterized by:

– an application point: the intersection between the radial direction and the trace of the caisson;

– a direction: always radial;

– a magnitude: the distance between the application point and the trace of the results;

– a sense: positive (outside the trace of the caisson), negative (inside).

All the results are normalized with respect to the maximum value (X_{max}, given in the figure).

The distribution of normal pressures at different depths of the caisson along the outside wall of the shaft, are presented in Figure 1.30a after an imposed displacement of 0.5 mm. Normal pressures indeed increase with depth, and are higher in front of the caisson, because of the lateral loading. The lateral displacement also induces a gap behind the caisson, denoted by the zero lateral pressures at depths of 0.25 and 1.25 m.

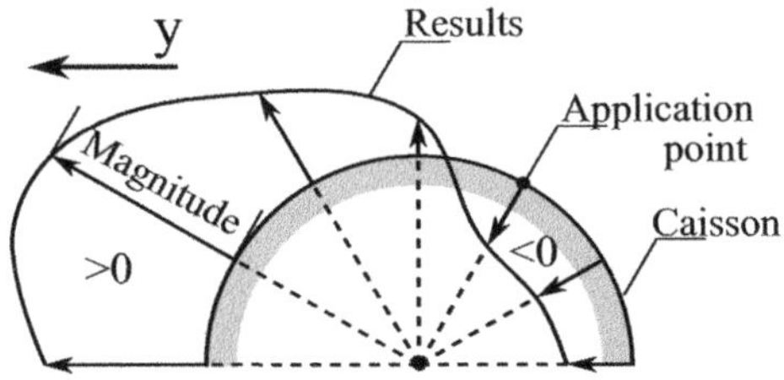

Figure 1.29. *Normalized radial distribution of the results*

The lateral movement induces horizontal shear stress along the outside wall of the shaft, which is always oriented tangentially to the shaft. Projections of these shear stresses in the y-direction are presented in Figure 1.31a. The maximum is obviously located on the side of the caisson where the tangent to the caisson is in the y-direction. This indicates that friction along the shaft also plays a role in balancing the applied lateral load.

Vertical shear stresses along the outside wall of the shaft are presented in Figure 1.32. The sign of these vertical shear stresses change between the front and the back of the caisson due to the rotational movement. They are of higher magnitude at the front since the increasing normal pressure allows a higher magnitude of shear stress. The axis of rotation of the caisson is perpendicular to the plane of symmetry and has its origin almost at the center of the caisson. It crosses the shaft of the caisson where the sign of the shear stress changes.

Failure can be defined after an imposed lateral displacement of 2.5 mm, where a slope breakage occurs in the load–displacement relationship. Normal pressures and horizontal shear stresses are presented at this moment in Figures 1.30b and 1.31b respectively.

The gap is completely open all along the back part of the shaft, and the pressure is equal to zero along this part. However, they are increased by almost 50% at the front. Once again, the maximum pressures increase with depth, due to the confinement. Shear stress in the y-direction have also increased, but the maximum has moved to the front of the caisson. This is due to the gap opening that has decreased the pressure and then the maximum shear stress available. In addition, it must be kept in mind that there is an interaction between the two shear stress components since

$$\tau_{max} = \mu \, p'_N = \sqrt{(\tau_1)^2 + (\tau_2)^2}. \quad [1.39]$$

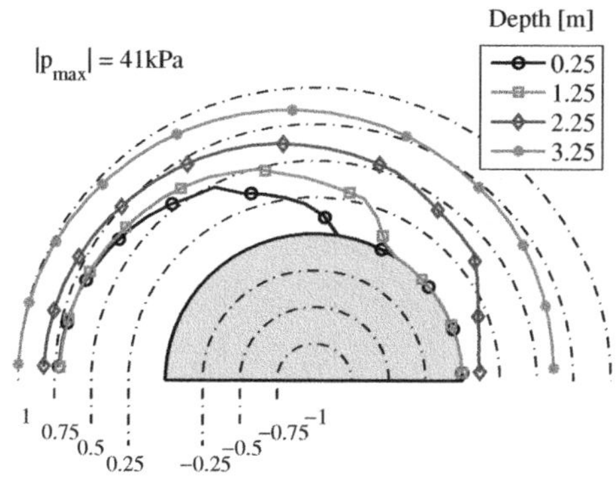

(a) Imposed lateral displacement equal to 0.5 mm

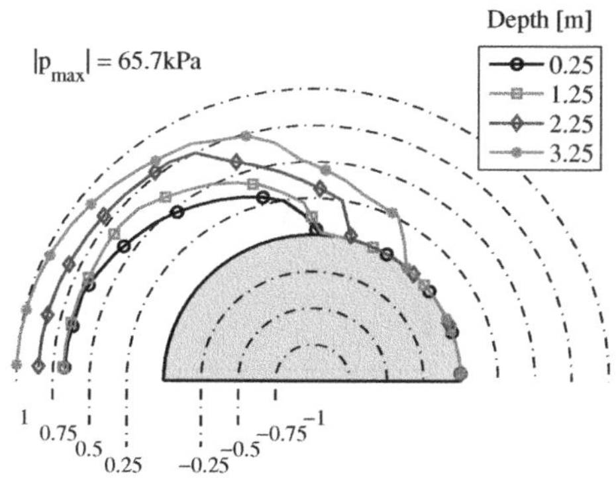

(b) Imposed lateral displacement equal to 2.5 mm

Figure 1.30. *Distribution of normal effective pressure along the outer shaft: lateral drained simulation*

Therefore, if rotation increases and mobilizes too much vertical shear stresses, then the lateral shear stress will also be reduced.

P-y curves are a useful tool to understand the basic working of piles that are laterally loaded and to simulate it as a 1D problem. They provide an insight into the stiffness, the strength and the displacement history of the caisson at different depths. Basically, the lateral loads acting on the caisson over a given range of depths are summed up and expressed as a force per unit length of foundation p. They are shown as a function of the lateral displacement of the neutral axis of the foundation at the same depth y. The total load incorporates contact pressure, lateral shear stress and pore water pressures. These curves are presented in Figure 1.33 at four depths.

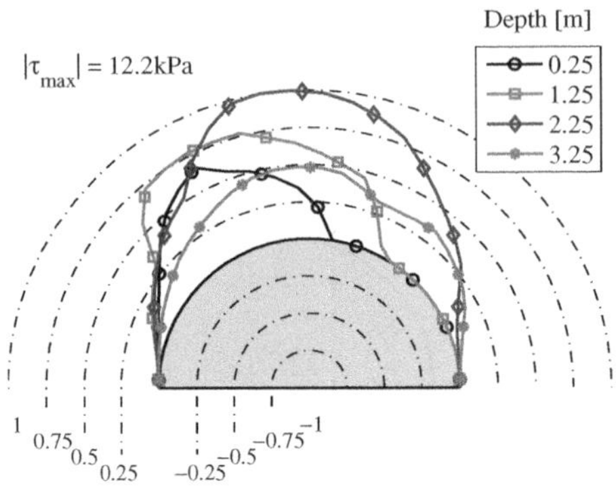

(a) Imposed lateral displacement equal to 0.5 mm

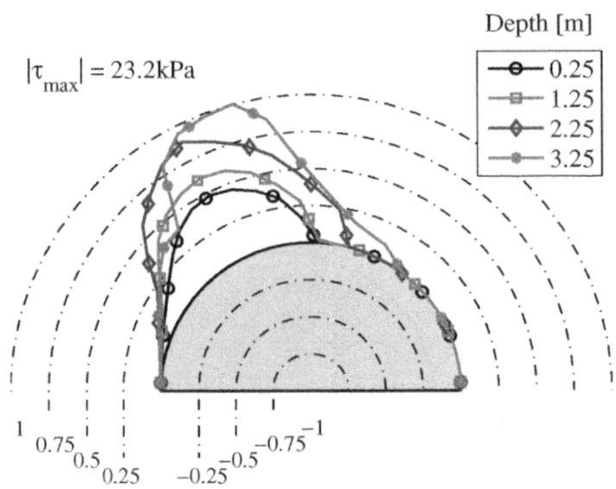

(b) Imposed lateral displacement equal to 2.5 mm

Figure 1.31. *Distribution of horizontal shear stress along the outer shaft (projection into the y-direction): lateral drained simulation*

However, a caisson is far from a pile: it has a length-to-diameter ratio of 0.5. There is then almost no flexion and the caisson cannot be modeled as a beam. Moreover, loads acting on the shaft are both inside and outside the caisson. Therefore, the p-y curves are the results of the difference between inside and outside reactions from the soil onto the caisson, in the y-direction. The y displacement is the displacement of the mean fiber of the caisson at each depth.

As expected, the lateral load increases while the depth decreases, which is at the origin of the rotational movement of the caisson. The lowest lateral displacement occurs at a higher depth. Each p-y curve also exhibits a sharp slope breakage already observed in the global displacement load curve. However, the lower the depth, the sooner the slope breakage occurs.

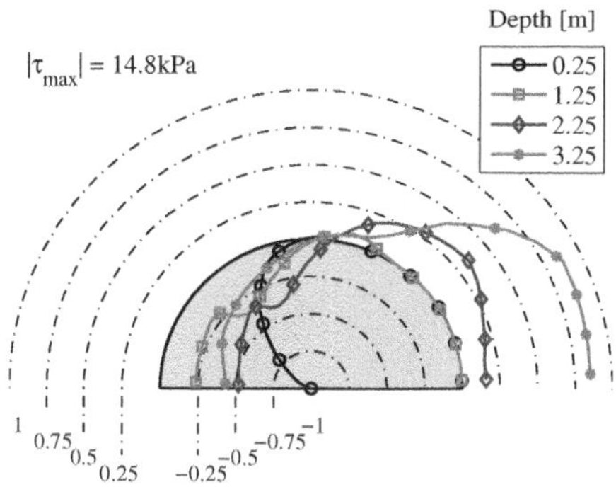

Figure 1.32. *Distribution of vertical shear stress along the outer shaft: lateral drained simulation, imposed lateral displacement equal to 0.5 mm*

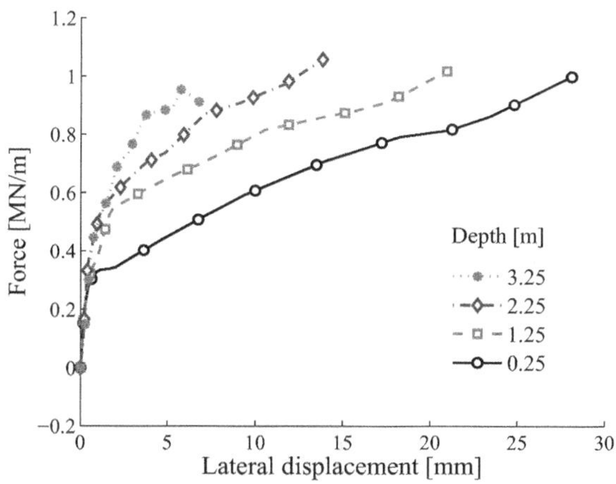

Figure 1.33. *P-y curves along the shaft of the caisson: drained simulation*

Similar to the tension simulation, the interface element is essential to reproduce the movement of the caisson. Indeed, rotation and gap opening, resulting from the lateral movement, can be only reproduced by interface elements. Additional features induced by the hydro-mechanical couplings are detailed in the following section.

1.4.4.2. *Partially drained simulation (coupled problem)*

Similar to the tension loading simulation, the partially drained nature of the caisson's behavior increases its ductility and strength. The load–displacement relationships are presented in Figure 1.34 in partially drained conditions. The total load drained curve is added for comparison. The difference in resistance after an imposed lateral displacement of 20 mm is almost 1 MN. Both vertical shear components ΔF_{in} and ΔF_{out} decrease with respect to the drained simulation, indicating a modification of shear distribution along the shaft. Finally, the uplifting movement of the caisson creates a gap under its top surface. Subsequently, a pore pressure drop occurs and the ΔF_{pw} is not null. This load opposes to the rotational movement of the caisson.

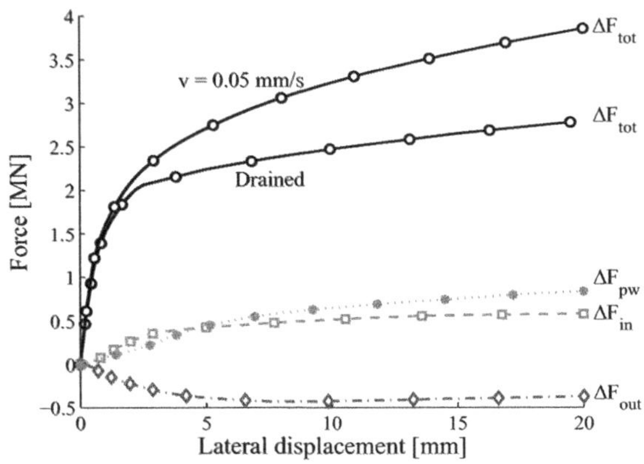

Figure 1.34. *Partially drained lateral simulation of the caisson (v = 0.05 mm/s): components of reaction for the entire caisson*

The distribution of pore water pressures is no longer symmetric, as shown in Figure 1.35. Variations of pore water pressures within the caisson are negative due to the slight uplifting movement. In this figure, they are more negative in the left corner of the caisson because a gap is also opening vertically in this zone, due to the lateral movement.

In front of the caisson, water overpressures should be generated since the soil is under compression. However, these pressures are quickly dissipated, since the material is very close to the mudline, where pressures are fixed. On the contrary, at the back, very negative pressures are generated, close to the tip of the caisson. It is thought that this is due to the longitudinal gap opening behind the caisson. On the one hand, the gap opening induces a suction effect and a drop of pressure. On the other hand, this gap increases the vertical fluid flow along the shaft and should decrease the suction effect. In this case, the first effect seems to be of greater magnitude than the second effect, but this depends on many variables such as soil permeability/stiffness and loading rate.

The drained and partially drained p-y curves are presented in Figure 1.36 for comparison. Each p-y curve includes different components acting on each side of the caisson: normal effective pressure, horizontal shear stress and pore water pressure. The curves close to the surface (depth 0.25 m) have more or less an identical shape. All the other partially drained curves show an increase in force for an identical displacement.

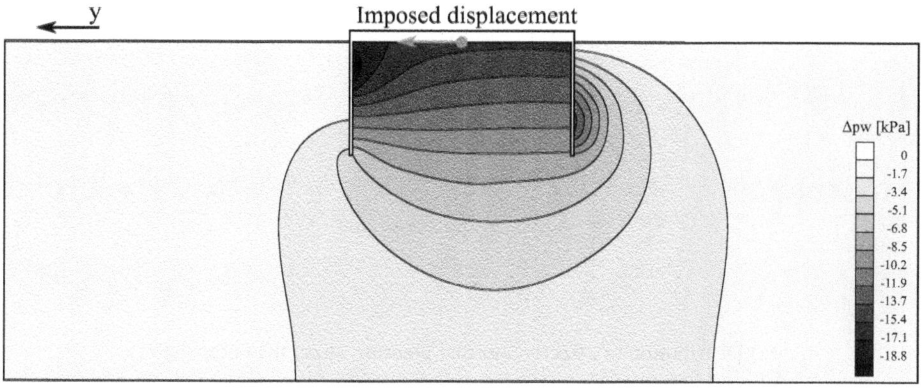

Figure 1.35. *Partially drained lateral displacement of the caisson after an imposed lateral displacement of 20 mm: variation of pore water pressure (Δp_w), cross-section in the plane Y-Z, drawing in the initial configuration*

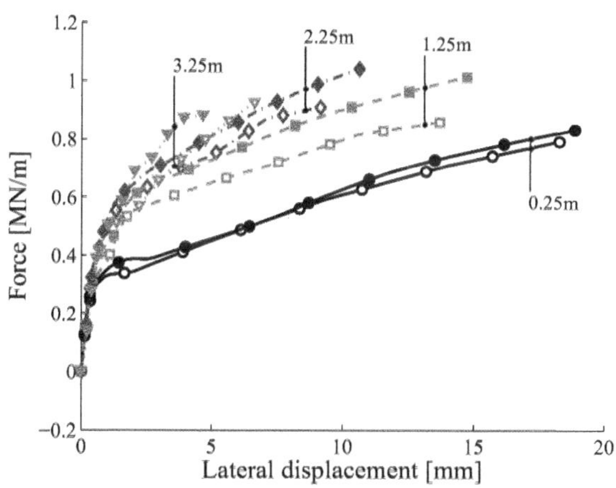

Figure 1.36. *P-y curves along the shaft of the caisson: partially drained simulation (v=0.05 mm/s)*

The distribution of effective stress around the shaft is not really modified, as shown in Figure 1.37a. Only the gap is less open, since the normal pressure is not equal to zero behind the caisson (depth 3.25 m). Therefore, it could be reasonably stated that the difference between p-y curves is due to the pore pressure distribution.

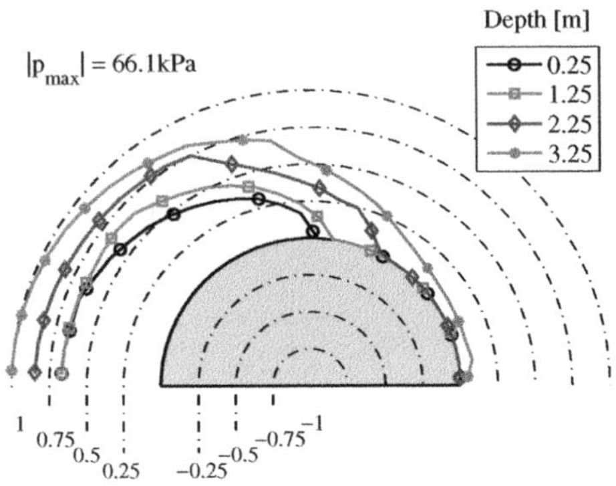

(a) Distribution of effective normal pressure along the outer shaft

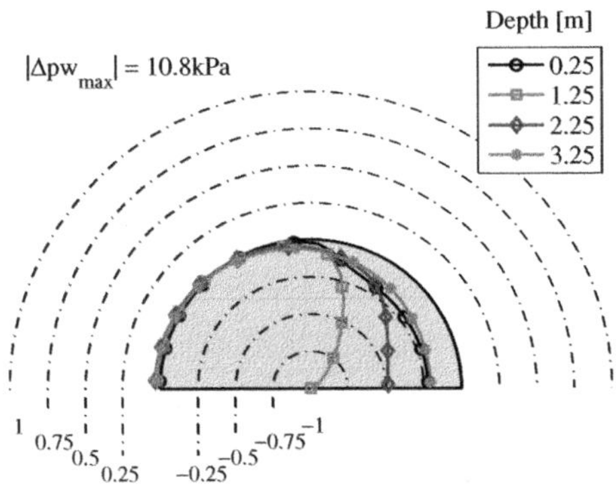

(b) Distribution of pore water pressure variations along the outer shaft

Figure 1.37. *Lateral partially drained simulation, with an imposed lateral displacement equal to 2.5 mm*

Variations of pore water pressure with respect to the hydrostatic distribution Δp_w within interface elements outside the caisson are presented in Figure 1.37b at different depths. The absolute magnitude of pressure variations is lower within the interface than within the soil because its permeability limits the transversal flow to the interface. There are almost no overpressures in front of the caisson, where compression of the soil holds. This is thought to be due to the combination of high sand permeability and a short drainage path close to the mudline.

Figure 1.38 presents the influence of the displacement rate on the load–displacement results. The strength and ductility increase with the increasing rate of loading. However, this increase is less important than when it is observed in tension simulations. It is not linear, and a very high rate does not involve a very large increase in loading. Indeed, in laterally loaded cases, negative variations of pore pressures are generated outside the caisson and dissipate much faster than in the tension case.

Results of simulations where the friction coefficient is equal to zero are also presented in Figure 1.38. This clearly reduces the resistance to lateral loading. This reduction arises from the loss of the horizontal shear stress component along the shaft, which directly opposes the loading. In addition, vertical shear stress no longer balances the rotational movement.

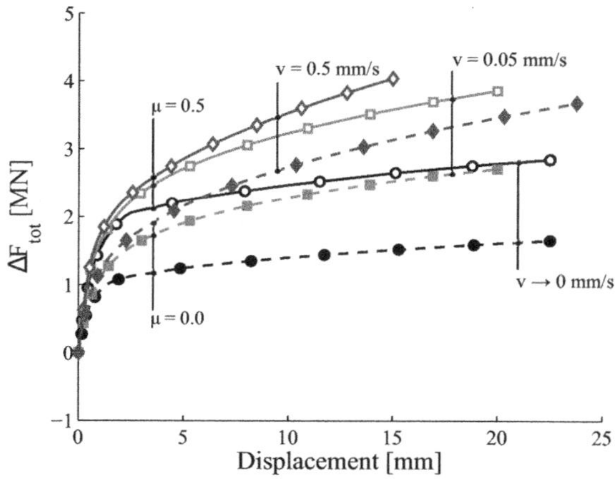

Figure 1.38. *Partially drained lateral simulation of the caisson: influence of the loading rate (0 is for a drained simulation) and friction (μ coefficient equal to 0 or 0.5)*

1.5. Conclusion and perspectives

1.5.1. *Conclusion*

The objectives of this chapter are to describe the formulation of a hydro-mechanically coupled finite element of interface and to illustrate its importance in the simulation of the soil–structure interaction in offshore engineering.

The main developments of a zero-thickness element are provided in sections 1.2 and 1.3. The contact constraint is ensured by a penalty method and computed on a segment-to-segment basis. The flow discretization is taken into account by a three-node discretization method, namely, the flow inside the interface is discretized separately from the flow inside volume elements. Three hydro-mechanical couplings are introduced:

– definition of a total pressure (p_N) as the sum of effective contact (p'_N) and pore water (p_w) pressures;

– storage component due to gap opening ($g_N > 0$);

– definition of fluid flow along the interface according to the cubic law.

A suction caisson typical of foundations for offshore wind turbines is further investigated on pure tension and lateral loading. In both cases, the soil–structure interactions are highly influenced by interface behaviors. Friction mobilization is crucial in the case of pure traction, and is also involved in the resistance to lateral loading. The gap creation under the lid and along the shaft of the caisson is also quite important, since it decreases the contact zone and the possible mobilization of shear stress. For all of these reasons, the use of a finite element of interface ruling contact detection and friction mobilization is necessary.

Behavior of suction caissons is neither purely drained nor purely undrained. Pore water pressures are generated and dissipated progressively during the loading of the caisson. This partially drained behavior requires hydro-mechanically coupled simulations. The coupled finite element of interface is developed for that purpose. The results indicate that the strength of caissons to both tension and lateral loading is increased by partially drained effects. In addition, the generation of increased or decreased pore water pressure is a reversible process, while variations of effective stress within the soil generate plastic strain. Therefore, the hydro-mechanical couplings can help reduce the settlement of the caisson.

Movement of the caisson in both directions involves gap opening between the lid, the shaft or both and the soil. In turn, negative variations of pore water pressures are generated, which involves a suction effect, balancing the loading. This has a stronger impact on tension than on lateral loading, but is significant in both cases.

1.5.2. *Perspectives*

A basic interface element was provided in this chapter and its importance was demonstrated. It is easy to implement and use, but is not free from numerical issues. Pure contact mechanics is a very dynamic field of research that provides new methods to avoid oscillations in pressure distribution or discretization issues. The interested reader should refer to [WRI 06] for an excellent summary of numerical methods available.

Similarly, a basic constitutive law is described, but more advanced models (e.g. dilatancy, cyclic degradation) are available, as described in the introduction. The literature reporting these developments is mostly concentrated on material behavior, and rarely on applications in offshore engineering. The main challenge regarding complex constitutive laws is the multiplication of material parameters that must be calibrated while *in situ* and experimental results are, most of the time, limited.

The fast development of offshore foundation technologies (e.g. very large piles, new anchor types) and the increasing complexity of simulations are likely to generalize the use of coupled interface elements. Therefore, it is necessary to popularize these elements and provide guidelines for the choice of numerical parameters.

1.6. References

[ACH 10] ACHMUS M., THIEKEN K., "On the behavior of piles in non-cohesive soil under combined horizontal and vertical loading", *Acta Geotechnica*, vol. 5, no. 3, pp. 199–210, 2010.

[ACH 13] ACHMUS M., AKDAG C.T., THIEKEN K., "Load-bearing behavior of suction bucket foundations in sand", *Applied Ocean Research*, vol. 43, pp. 157–165, 2013.

[ALO 13] ALONSO E., ZANDARÍN M., OLIVELLA S., "Joints in unsaturated rocks: Thermo-hydro-mechanical formulation and constitutive behaviour", *Journal of Rock Mechanics and Geotechnical Engineering*, vol. 5, no. 3, pp. 200–213, Elsevier, 2013.

[AND 08] ANDERSEN K., JOSTAD H., DYVIK R., "Penetration resistance of offshore skirted foundations and anchors in dense sand", *Journal of Geotechnical and Geoenvironmental Engineering*, vol. 134, pp. 106–116, 2008.

[BAR 98] BARNICHON J., Finite element modelling in structural and petroleum geology, PhD thesis, University of Liège, Belgium, 1998.

[BEL 91] BELYTSCHKO T., NEAL M., "Contact impact by the pinball algorithm with penalty and Lagrangian methods", *International Journal for Numerical Methods in Engineering*, vol. 31, pp. 547–572, 1991.

[BEL 98] BELGACEM F., HILD P., LABORDE P., "The mortar finite element method for contact problems", *Mathematical and Computer Modelling*, vol. 28, no. 4, pp. 263–271, Elsevier, 1998.

[BER 88] BERKOWITZ B., BEAR J., BRAESTER C., "Continuum models for contaminant transport in fractured porous formations", *Water Resources Research*, vol. 24, no. 8, pp. 1225–1236, 1988.

[BIE 12] BIENEN B., DÜHRKOP J., GRABE J., RANDOLPH M., WHITE D., "Response of piles with wings to monotonic and cyclic lateral loading in sand", *Journal of Geotechnical and Geoenvironmental Engineering*, vol. 138, no. 3, pp. 364–375, 2012.

[BIE 18a] BIENEN B., KLINKVORT R.T., O'LOUGHLIN C., ZHU F., BYRNE B.W., "Suction caissons in dense sand, Part I: Installation, limiting capacity and drainage", *Géotechnique*, vol. 11, pp. 1–47, 2018.

[BIE 18b] BIENEN B., KLINKVORT R.T., O'LOUGHLIN C.D., ZHU F., BYRNE B.W., "Suction caissons in dense sand, Part II: Vertical cyclic loading into tension", *Géotechnique*, vol. 68, no. 11, pp. 1–15, 2018.

[BOR 13] BORJA R.I., *Plasticity: Modeling & Computation*, Springer Science & Business Media, Berlin, 2013.

[BOU 68] BOUSSINESQ J., "Mémoire sur l'influence des frottements dans les mouvements réguliers des fluides", *Journal de Mathématiques Pures et Appliquées*, 1868.

[BRA 09] BRANSBY M.F., YUN G.-J., "The undrained capacity of skirted strip foundations under combined loading", *Géotechnique*, vol. 59, no. 2, pp. 115–125, 2009.

[BYR 02] BYRNE B., HOULSBY G., "Experimental investigations of response of suction caissons to transient vertical loading", *Journal of the Geotechnical and Geoenvironmental Engineering*, vol. 128, no. 11, pp. 926–939, 2002.

[BYR 03] BYRNE B.W., HOULSBY G.T., "Foundations for offshore wind turbines", *Philosophical Transactions of the Royal Society A: Mathematical, Physical and Engineering Sciences*, vol. 361, no. 1813, pp. 2909–2930, 2003.

[CER 15] CERFONTAINE B., DIEUDONNÉ A., RADU J., COLLIN F., CHARLIER R., "3D zero-thickness coupled interface finite element: Formulation and application", *Computers and Geotechnics*, vol. 69, pp. 124–140, 2015.

[CER 16] CERFONTAINE B., COLLIN F., CHARLIER R., "Numerical modelling of transient cyclic vertical loading of suction caissons in sand", *Geotechnique*, vol. 66, no. 2, pp. 121–136, 2016.

[CHA 88] CHARLIER R., CESCOTTO S., "Modélisation du phénomène de contact unilatéral avec frottement dans un contexte de grandes déformations", *Journal de Mécanique Théorique et Appliquée*, vol. 7, no. 1, Gauthier-Villars, 1988.

[DAY 94] DAY R., POTTS D., "Zero thickness interface elements – Numerical stability and application", *International Journal for Numerical and Analytical Methods in Geomechanics*, vol. 18, pp. 689–708, 1994.

[DEJ 06] DEJONG J., WHITE D., RANDOPH M., "Microscale observation and modeling of soil structure interface behavior using particle image velocimetry", *Soils and Foundations*, vol. 46, no. 1, pp. 15–28, 2006.

[DES 84] DESAI C., ZAMAN M., LIGHTNER J., SIRIWARDANE H., "Thin-layer element for interfaces and joints", *International Journal for Numerical and Analytical Methods in Geomechanics*, vol. 8, no. 1, pp. 19–43, 1984.

[FIS 06] FISCHER K., WRIGGERS P., "Mortar based frictional contact formulation for higher order interpolations using the moving friction cone", *Computer Methods in Applied Mechanics and Engineering*, vol. 195, no. 37, pp. 5020–5036, Elsevier, 2006.

[GAO 13] GAO Y., QIU Y., LI B., LI D., SHA C., ZHENG X., "Experimental studies on the anti-uplift behavior of the suction caissons in sand", *Applied Ocean Research*, vol. 43, pp. 37–45, 2013.

[GEN 90] GENS A., CAROL I., ALONSO E., "A constitutive model for rock joints formulation and numerical implementations", *Computers and Geotechnics*, vol. 9, pp. 3–20, 1990.

[GOO 68] GOODMAN R., TAYLOR R., BREKKE T., "A model for the mechanics of jointed rock", *Journal of the Soil Mechanics and Foundations Division*, vol. 94, 1968.

[GOU 08] GOURVENEC S., "Effect of embedment on the undrained capacity of shallow foundations under general loading", *Géotechnique*, vol. 58, no. 3, pp. 177–185, 2008.

[GOU 09] GOURVENEC S., ACOSTA-MARTINEZ H., RANDOLPH M., "Experimental study of uplift resistance of shallow skirted foundations in clay under transient and sustained concentric loading", *Géotechnique*, vol. 59, no. 6, pp. 525–537, 2009.

[GUI 02] GUIDUCCI C., PELLEGRINO A., RADU J.-P., COLLIN F., CHARLIER R., "Numerical modeling of hydro-mechanical fracture behavior", *Numerical Models in Geomechanics*, pp. 293–299, 2002.

[HAB 98] HABRAKEN A., CESCOTTO S., BANNING Q., "Contact between deformable solids: The fully coupled approach", *Mathematical and Computer Modelling*, vol. 28, no. 4, pp. 153–169, 1998.

[HAL 85] HALLQUIST J., GOUDREAU G., BENSON D., "Sliding interfaces with contact-impact in large-scale Lagrangian computations", *Computer Methods in Applied Mechanics and Engineering*, vol. 51, no. 1, pp. 107–137, Elsevier, 1985.

[HAR 09] HARTMANN S., OLIVER J., WEYLER R., CANTE J., HERNÁNDEZ J., "A contact domain method for large deformation frictional contact problems. Part 2: Numerical aspects", *Computer Methods in Applied Mechanics and Engineering*, vol. 198, no. 33, pp. 2607–2631, Elsevier, 2009.

[HO 11] HO T., JARDINE R., ANH-MINH N., "Large-displacement interface shear between steel and granular media", *Géotechnique*, vol. 61, no. 3, pp. 221–234, 2011.

[HOU 05a] HOULSBY G., BYRNE B., "Design procedures for installation of suction caissons in sand", *Proceedings of the Institution of Civil Engineers – Geotechnical Engineering*, vol. 158, no. 3, pp. 135–144, 2005.

[HOU 05b] HOULSBY G., IBSEN L., BYRNE B., "Suction caissons for wind turbines", *International Symposium on Frontiers in Offshore Geotechnics*, pp. 75–93, 2005.

[HOU 05c] HOULSBY G., KELLY R., BYRNE B., "The tensile capacity of suction caissons in sand under rapid loading", *Frontiers in Offshore Geotechnics*, no. 1, pp. 405–410, 2005.

[HOU 06] HOULSBY G., KELLY R., HUXTABLE J., BYRNE B., "Field trials of suction caissons in sand for offshore wind turbine foundations", *Géotechnique*, vol. 56, no. 1, pp. 3–10, 2006.

[HU 04] HU L., PU J., "Testing and modeling of soil-structure interface", *Journal of Geotechnical and Geoenvironmental Engineering*, vol. 130, no. 8, pp. 851–860, 2004.

[IBS 14] IBSEN L.B., LARSEN K.A., BARARI A., "Calibration of failure criteria for bucket foundations on drained sand under general loading", *Journal of Geotechnical and Geoenvironmental Engineering*, vol. 140, no. 7, 2014.

[ISK 02] ISKANDER M., EL-GHARBAWY S., OLSON R., "Performance of suction caissons in sand and clay", *Canadian Geotechnical Journal*, vol. 39, no. 3, pp. 576–584, 2002.

[JHA 14] JHA B., JUANES R., "Coupled multiphase flow and poromechanics: A computational model of pore pressure effects on fault slip and earthquake triggering", *Water Resources Research*, vol. 50, pp. 3376–3808, 2014.

[JOH 92] JOHNSON K., *Contact Mechanics*, Cambridge University Press, New York, 1992.

[KEL 06] KELLY R., HOULSBY G., BYRNE B., "A comparison of field and laboratory tests of caisson foundations in sand and clay", *Géotechnique*, vol. 56, no. 9, pp. 617–626, 2006.

[KLA 88] KLARBRING A., BJÖRKMAN G., "A mathematical programming approach to contact problems with friction and varying contact surface", *Computers & Structures*, vol. 30, no. 5, pp. 1185–1198, 1988.

[KOU 14] KOURKOULIS R., LEKKAKUS P., GELAGOTI F., KAYNIA A., "Suction caisson foundations for offshore wind turbines subjected to wave and earthquake loading: Effect of soil–foundation interface", *Géotechnique*, vol. 64, no. 3, pp. 171–185, 2014.

[LEW 98] LEWIS R., SCHREFLER B., *The Finite Element Method in the Static and Dynamic Deformation and Consolidation of Porous Media*, John Wiley & Sons, New York, 1998.

[LI 15] LI D., ZHANG Y., FENG L., GAO Y., "Capacity of modified suction caissons in marine sand under static horizontal loading", *Ocean Engineering*, vol. 102, pp. 1–16, 2015.

[LIU 06] LIU H., SONG E., LING H., "Constitutive modeling of soil-structure interface through the concept of critical state soil mechanics", *Mechanics Research Communications*, vol. 33, pp. 515–531, 2006.

[LIU 08] LIU H., LING H., "Constitutive description of interface behavior including cyclic loading and particle breakage within the framework of critical state soil mechanics", *International Journal for Numerical and Analytical Methods in Geomechanics*, vol. 32, no. 12, pp. 189–213, 2008.

[MCC 56] MCCLELLAND B. *et al.*, "Soil modulus for laterally loaded piles", *Journal of the Soil Mechanics and Foundations Division*, vol. 82, no. 4, pp. 1–22, 1956.

[MOR 02] MORTARA G., BOULON M., GHIONNA V., "A 2-D constitutive model for cyclic interface behaviour", *International Journal for Numerical and Analytical Methods in Geomechanics*, vol. 26, no. 11, pp. 1071–1096, 2002.

[NG 97] NG K., SMALL J., "Behavior of joints and interfaces subjected to water pressure", *Computers and Geotechnics*, vol. 20, no. 1, pp. 71–93, 1997.

[OLI 09] OLIVER J., HARTMANN S., CANTE J., WEYLER R., HERNÁNDEZ J., "A contact domain method for large deformation frictional contact problems. Part 1: Theoretical basis", *Computer Methods in Applied Mechanics and Engineering*, vol. 198, no. 33, pp. 2591–2606, Elsevier, 2009.

[OLS 01] OLSSON R., BARTON N., "An improved model for hydromechanical coupling during shearing of rock joints", *International Journal of Rock Mechanics and Mining Sciences*, vol. 38, no. 3, pp. 317–329, 2001.

[ORO 98] ORON A., BERKOWITZ B., "Flow in rock fractures: The local cubic law assumption reexamined", *Water Resources Research*, vol. 34, no. 11, pp. 2811–2825, 1998.

[PUS 04] PUSO M., LAURSEN T., "A mortar segment-to-segment contact method for large deformation solid mechanics", *Computer Methods in Applied Mechanics and Engineering*, vol. 193, no. 6, pp. 601–629, 2004.

[RAG 19] RAGNI R., BIENEN B., STANIER S., O'LOUGHLIN C., CASSIDY M., "Observations during suction bucket installation in sand", *International Journal of Physical Modelling in Geotechnics*, pp. 1–49, 2019.

[REE 10] REESE L., VAN IMPE W., *Single Piles and Pile Groups under Lateral Loading*, 2nd edition, CRC Press, Boca Raton, 2010.

[SEG 08a] SEGURA J., CAROL I., "Coupled HM analysis using zero-thickness interface elements with double nodes. Part I: Theoretical model", *International Journal for Numerical and Analytical Methods in Geomechanics*, vol. 32, no. 18, pp. 2083–2101, 2008.

[SEG 08b] SEGURA J., CAROL I., "Coupled HM analysis using zero-thickness interface elements with double nodes. Part II: Verification and application", *International Journal for Numerical and Analytical Methods in Geomechanics*, vol. 32, no. 18, pp. 2103–2123, 2008.

[SEN 82] SENPERE D., AUVERGNE G., "Suction anchor piles – A proven alternative to driving or drilling", *Proceedings of the 14th Offshore Technology Conference*, pp. 486–493, 1982.

[SEN 08] SENDERS M., Suction caissons in sand as tripod foundations for offshore wind turbines, PhD thesis, University of Western Australia, 2008.

[SEN 09] SENDERS M., RANDOLPH M., "CPT-based method for the installation of suction caissons in sand", *Journal of Geotechnical and Geoenvironmental Engineering*, vol. 135, pp. 14–25, 2009.

[SHA 93] SHARMA K., DESAI C., "Analysis and implementation of thin-layer element for interfaces and joints", *Journal of Engineering Mechanics*, vol. 118, no. 12, pp. 2442–2462, 1993.

[SHA 97] SHAHROUR I., REZAIE F., "An elastoplastic constitutive relation for the soil-structure interface under cyclic loading", *Computers and Geotechnics*, vol. 21, no. 1, pp. 21–39, 1997.

[STU 16] STUTZ H., MASIN D., "Hypoplastic interface models for fine-grained soils", *International Journal for Numerical and Analytical Methods in Geomechanics*, vol. 41, no. 2, pp. 284–303, 2016.

[TER 25] TERZAGHI K., *Erdbaumechanik auf Bodenphysikalischer Grundlage* [The Mechanics of Earth Construction Based on Soil Physics], Deuticke, Leipzig, 1925.

[THI 14] THIEKEN K., ACHMUS M., SCHRÖDER C., "On the behavior of suction buckets in sand under tensile loads", *Computers and Geotechnics*, vol. 60, pp. 88–100, 2014.

[TJE 90] TJELTA T., AAS P., HERMSTAD J., ANDENAES E., "The skirt piled Gullfaks C platform installation", *Proceedings of the Offshore Technology Conference*, 1990.

[TRA 05] TRAN M., Installation of suction caissons in dense sand and the influence of silt and cemented layers, PhD thesis, The University of Sydney, 2005.

[TSA 81] TSANG Y., WITHERSPOON P., "Hydromechanical behavior of a deformable rock fracture subject to normal stress", *Journal of Geophysical Research: Solid Earth*, vol. 86, no. B10, pp. 9287–9298, 1981.

[WEI 15] WEISSENFELS C., WRIGGERS P., "A contact layer element for large deformations", *Computational Mechanics*, vol. 55, no. 5, pp. 873–885, 2015.

[WIT 80] WITHERSPOON P., WANG J., IWAI K., GALE J., "Validity of cubic law for fluid flow in a deformable rock fracture", *Water Resources Research*, vol. 16, no. 6, pp. 1016–1024, 1980.

[WRI 01] WRIGGERS P., KRSTULOVIC-OPARA L., KORELC J., "Smooth C1-interpolations for two-dimensional frictional contact problems", *International Journal for Numerical Methods in Engineering*, vol. 51, no. 12, pp. 1469–1495, 2001.

[WRI 04] WRIGGERS P., ZAVARISE G., "Computational contact mechanics", in STEIN E., DE BORST R., HUGHES T. (eds), *Encyclopedia of Computational Mechanics, Volume 2: Solids and Structures*, pp. 195–226, John Wiley & Sons, Chichester, 2004.

[WRI 06] WRIGGERS P., *Computational Contact Mechanics*, 2nd edition, John Wiley & Sons, Chichester, 2006.

[WRI 13] WRIGGERS P., SCHRÖDER J., SCHWARZ A., "A finite element method for contact using a third medium", *Computational Mechanics*, vol. 52, no. 4, pp. 837–847, 2013.

[ZAN 13] ZANDARIN M., ALONSO E., OLIVELLA S., "A constitutive law for rock joints considering the effects of suction and roughness on strength parameters", *International Journal of Rock Mechanics and Mining Sciences*, vol. 60, pp. 333–344, Elsevier, 2013.

[ZIE 00] ZIENKIEWICZ O.C., TAYLOR R.L., *Finite Element Method: Volume 1*, 5th edition, Butterworth-Heinemann, Oxford, 2000.

2

DEM Approach of the Modeling for Geotechnical Structures in Interaction with Reinforcements

The sophistication of geotechnical structures (new technologies and innovative materials in civil engineering) implies the use of increasingly sophisticated numerical models for their design. The ever-increasing improvement in the computational capabilities, and the emergence of new methods of calculation, make possible now what was impossible to envisage a few decades ago, particularly in terms of duration and number of elements of the numerical simulations. Among the numerical methods in full development are discrete methods. Although often used for research purposes, these methods are increasingly used in engineering, in view of the certain advantages that they represent.

2.1. Introduction

The SSI problem addressed in this chapter is studied using the DEM (Discrete Element Method). The basic principle of discrete methods is to consider a material as a set of particles that interact at their contact points. Therefore, these methods ideally apply to granular materials due to their discrete nature and make it possible – considering a limited number of parameters – to easily reproduce the behavior of soil (load transfer mechanisms, arching effects, expansion, cracking or collapse). These methods apply to quasistatic problems and see their main interest in problems that involve large deformations or collapsed areas, as well as to cyclic or dynamic applications, easily modeled given the specific mathematical formulation used in DEM. Note that the use of discrete element modeling does not necessarily imply

---

Chapter written by Pascal VILLARD.

having to faithfully represent either the real form or the number of grains present in a geo-material. In any case, this approach would not be feasible from the point of view of the present abilities of the computers. In fact, the shape, the distribution and the number of granular elements are chosen for the simple purpose of having a numerical material able to reproduce the macroscopic mechanical behavior of the granular material. While the DEM is highly relevant for the modeling of discrete materials such as soils, it is naturally less apt to reproduce the behavior of reinforcement elements that are, by nature, continuous. To overcome this problem, specific elements and contact law were implemented in DEM modeling to restore the behavior of deformable reinforcements, and to take account of a realistic description of the interface behavior between the soil and the reinforcement.

After presenting briefly the discrete element method and the theoretical concepts used to take account of the soil–inclusion interaction, we will present different illustrations of this method and, in particular, the applications to geotechnical structures in interaction with rigid piles (quasistatic and cyclic loadings) or with flexible and deformable reinforcements (geosynthetic sheets).

2.2. Discrete modeling

2.2.1. *General concepts of the discrete modeling approach*

The various discrete methods differ mainly in the resolution scheme employed, the type of elements used, and the kind of interaction laws defined between the elements in contact. Two main families of methods are usually employed: the approaches resulting from contact dynamics and those derived from molecular dynamics. The main differences between these two methods result from the deformability tolerated or not between two elements at the contact points. In the contact dynamics approach, initiated by Moreau and Jean [JEA 92, 99], particles cannot interpenetrate and are considered non-deformable even at the contact points, so that contact interactions are controlled by shock laws, involving temporal discontinuities of velocities and forces.

The dynamic molecular method is also based on non-deformable elements but those which are able to overlap slightly at the contact points. Based on molecular dynamics, the DEM was first introduced by Cundall [CUN 71], for applications to rock mechanics problems, and then applied to civil engineering by Cundall and Strack [CUN 79]. Contact laws defined at the microscopic level make it possible to establish the interaction forces taking into account the overlapping and relative displacement between the two elements in contact. The numerical scheme used to solve the quasistatic or dynamic problem is an iterative process that successively alternates the resolution of Newton's second law of motion, applied to the discrete elements, and the actualization of the interaction forces at each contact point. The motion equations are integrated using an explicit centered finite difference algorithm, involving a time step Δt. This numerical scheme allows very large displacements, alternative movements between the elements, shocks and dynamics behaviors.

The basic elements used in the discrete element method are generally spherical in shape since these forms have the advantage of facilitating the contact detection process. Nevertheless, in an assembly of spheres, the rolling between particles is such that it is difficult to simulate a natural soil presenting a high friction angle value. In order to limit grain rotations, non-spherical elementary particles (polyhedron or sphero-polyhedron) or clumps (agglomerates of rigid and non-breakable spherical particles, which make it possible to optimize the contact detection) are most often used. In some cases, rolling resistance laws can be introduced to limit the excessive rotation of grains [IWA 98]. To avoid regular assemblies of particles that have singular behaviors, elements of different sizes are almost systematically introduced. The size distribution of elements, their shape and their arrangement have a preponderant role in the macroscopic behavior of the granular material [SAL 09, 11]. Particular attention must therefore be paid when placing the elements, in order to ensure the production of a homogeneous and isotropic assembly of particles at a fixed density. For the following applications to reinforced structures, and in order to realize numerical granular assemblies with various porosities, the particles were put in place using the ERDF (Expansion Radius and Decrease Friction [CHA 05]) methodology. This methodology makes it possible to produce homogeneous and isotropic particle assemblies.

Due to the fact that there is no explicit relation between the microscopic and macroscopic parameters (Young's modulus E, Poisson's ratio ν, peak friction angle φ_p and residual friction angle φ_r), a specific calibration methodology, based on the trial-and-error process, is required. A classical way to obtain the micromechanical parameters is to simulate laboratory tests, performed in well-controlled conditions (as triaxial tests, for example) and compare the numerical results to the experimental ones [SAL 09].

Classically, the interaction laws between two elements are most often expressed as a function of interpenetration and relative displacement between the particles, and are generally defined by a stiffness model, a slip model or possibly a cohesion model. Numerous contact models have been developed, ranging from the simplest based on linear elastic considerations, to the most advanced: nonlinear elasticity, elastoplasticity [LUD 05] and viscosity [IWA 00]. For the following applications to reinforced structures, a classical linear elastic contact law [DON 95], characterized by a friction coefficient δ and two normal and tangential contact stiffness coefficients

k_n and k_s, was used to define the contact behavior. The normal contact force can be written as a function of the overlap between two elements using equation [2.1]:

$$\{\vec{F}_n\} = k_n \{\vec{U}_n\} \qquad [2.1]$$

with $\{\vec{F}_n\}$ the normal contact force between the elements and $\{\vec{U}_n\}$ the overlap in the normal direction of the contact.

In the tangential direction, an incremental contact law, equation [2.2], is used to link the increase in tangential contact force to the increase of tangential relative displacement, calculated by considering the translation and the rotation of the two particles in contact.

$$d\{\vec{F}_t\} = k_s\, d\{\vec{U}_t\} \quad \text{with} \quad |\vec{F}_t| \le |\vec{F}_n|\, tg(\delta) \qquad [2.2]$$

with $\{\vec{F}_t\}$ the tangential contact force and $\{\vec{U}_t\}$ the tangential displacement.

The normal contact stiffness k_n (or the tangential contact stiffness k_s) between two spheres of radius R_i and R_j, expressed in N/m, is a function of the normal rigidity K_{nij} (or the tangential rigidity K_{sij}) of the two constitutive materials in contact, expressed in N/m^2, as defined in equation [2.3].

$$k_n = K_{nij}\,(R_i * R_j)/(R_i + R_j) \qquad [2.3]$$

Interaction laws, defined at a local scale, make it possible to reproduce the global macroscopic behavior of the particle assembly, considering various loading paths [SIB 19]. The macroscopic behavior of a set of particles strongly depends on the density of the granular medium, the shape of the particles and the value of the micromechanical contact parameters (normal and tangential stiffness, contact friction coefficient, grading, particle shape and density of the numerical sample). Usually, the macroscopic behavior restored leads to a granular material whose macroscopic friction angle differs from the value of the microscopic friction angle defined at the contact level, given the interlocking of the grains, which contributes to increasing its mechanical characteristics.

The stress tensor σ_{ij} within a volume V of the granular assembly can be computed by equation [2.4] considering the contact forces acting at all contact points included in the volume V [WEB 66]. N_c is the number of contact points in V, f^i is the projection of the contact force f on the i-axis and l^j is the projection of the branch vector l on the j-axis with i = x, y, z and j = x, y, z. The branch vector l is defined by the vector linking the mass centers of the elements in contact.

$$\sigma_{ij} = \frac{1}{V} \sum_{\alpha=1}^{N_c} f_\alpha^i l_\alpha^j \qquad [2.4]$$

2.2.2. *Specific interaction between discrete particles and reinforcement elements*

The interface between the reinforcement elements and the soil is the privileged area that allows the transmission of interaction forces from one element to another (a completely smooth and non-frictional interface induces non-transmission of frictional forces). The care taken in modeling the interface is therefore essential in hoping to highlight the real behavior of structures in reinforced soil. The use of discrete modeling makes it possible to very easily manage the interface behavior by introducing appropriate force–displacement interaction laws. There are two strategies for modeling boundary conditions or integrating reinforcement elements.

The first is to use a group of spheres associated together to form a wall or a reinforcing element that can deform according to the constitutive interaction laws defined between the spheres. The disadvantage of this method comes from the fact that there is a microroughness between the group of spheres and the particles of the soil that can be modified during the numerical simulation according to the extensibility of the group of spheres. It is therefore difficult to control the interface friction since uncontrolled structural friction is introduced during the simulation by the microroughness between the elements. Another important point lies in the fact that the discrete element method most often considers that the contact between two elements is lost as soon as the elements separate. It is therefore very difficult to ensure a regular transmission of the friction forces, from the particles of soil towards the reinforcement constituted by a sphere assembly, due to the contact microroughness that can involve a loss of contact when the relative displacements at the interface are very large.

The second strategy is to use plane walls as boundary conditions and to develop smooth specific elements to represent reinforcements as accurately as possible. This technique is the most relevant because it does not introduce structural friction (the macroscopic friction angle of the interface corresponds to the microscopic friction angle defined at the contact) and ensures the continuity of the contact even when the relative displacements at the interface are very large. This technique makes it possible to take account of very complex mechanisms acting at the interface as the rolling mechanisms of soil particles, or the expansion or decrease in volume due to the shearing forces.

Similar laws to those established between two basic particles, equations [2.1] and [2.2], may be used to define the interface behavior between a planar element (wall or reinforcement) and a soil particle. The normal stiffness k_{ni} at the contact prevents the interpenetration of the elements in contact, whereas the tangential stiffness k_{si} directly influences the rate of mobilization of the friction at the interface, for the case of small relative displacements and until the maximum value of the friction is obtained. The amplitude of the relative displacements at the interface and the contact tangential stiffness directly influence both the intensity and the orientation of the contact forces. The classical Mohr–Coulomb friction law, similar to those presented Figure 2.1, is usually used.

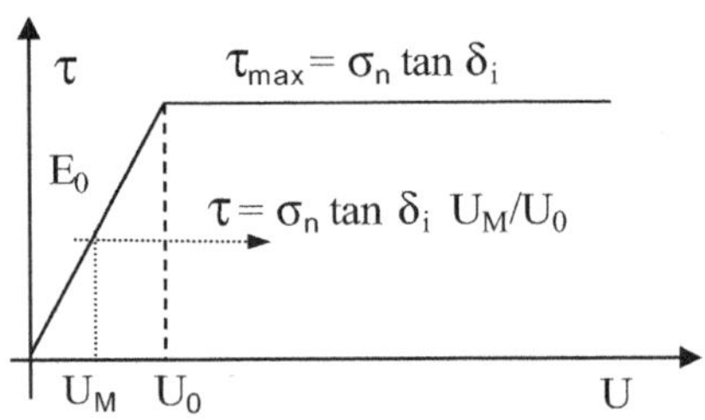

Figure 2.1. *Friction interface criterion*

τ and σ_n denote the tangential and normal stresses acting at the interface, U the relative displacement between soil particles and the planar element, U_0 the value of the minimal relative displacement necessary to mobilize the friction fully and E_0 the tangential stiffness modulus of the interface expressed in N/m^3 and defined by equation [2.5], where S represents the influence area of the contact ($S = \pi\, d^2/4$ for a spherical particle of diameter d).

$$E_0 = k_{si}/S \qquad [2.5]$$

Note that δ_i, the microscopic friction angle of the interface, is equal to the macroscopic friction angle that can be measured by a classical friction test. In practice, U_0 does not exceed a few millimeters. The intensity and the orientation of the friction force depend strongly on the value retained for U_0, especially for small relative displacements [VIL 09]. As an example, the friction response (blue lines) induced by the displacement along an elliptical path (red dashed line) of a spherical particle, in contact with a fixed plane, is presented in Figure 2.2. A vertical force keeps the contact persistent. We note that the smaller the value of U_0, the sooner full mobilization of the friction occurs so that the friction forces between the soil particle and the plane are increasingly tangential to the trajectory given to the sphere.

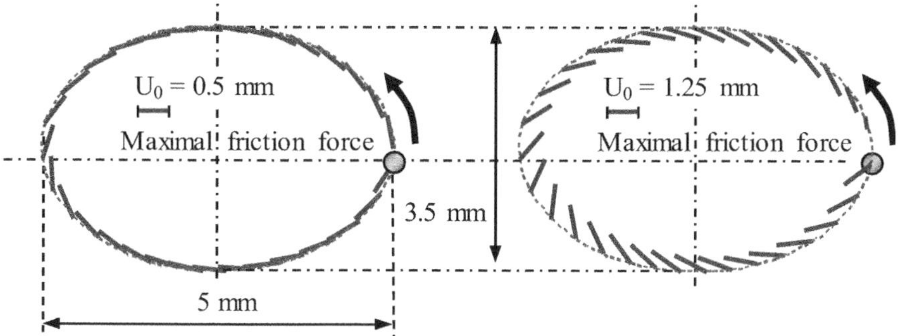

Figure 2.2. *Intensity and orientation of the tangential friction force for different values of U_0. For a color version of this figure, see www.iste.co.uk/grange/soil.zip*

2.2.3. *Numerical strategy for geotechnical structure modeling using DEM*

As mentioned, and due to the fact that a specific calibration methodology – based on trial and error – was used to calibrate the micromechanical parameters, the modeling of true-scale structures using DEM does not need to reproduce perfectly the grain morphology, the particle size distribution and the particle number, as this would imply a prohibitive calculation duration. However, to roughly mimic the shape of the soil particles and to approach the macroscopic mechanical behavior of real soil (particularly the internal friction angle at the peak) as well as possible, the granular material is modeled by means of an assembly of clusters of different sizes. The main results expected for applications to geotechnical structures are: the displacements of the discrete particles, the network of contacts within the granular assembly, the interaction forces between the soil and reinforcement, the stress tensor, the change in structure of the granular material and the mechanism of collapse.

2.3. Application of the DEM to geotechnical structures in interaction with rigid piles

Soil improvement by rigid inclusions is a technique used in many countries to limit surface settlements in highly compressible areas. This technique is based on the use of a network of concrete piles (assumed to be rigid due to their low deformability, compared to that of the compressible soil) and a granular load transfer layer positioned over the piles. This makes it possible to reduce the loads acting on the compressive subsoil by redirecting one part of the vertical loads related to the weight of the embankment and overloads to a rigid bedrock. There are two common applications of soft soil reinforcement, based on the use of a network of rigid piles: reinforcement under civil engineering buildings or embankments (road or railway embankments). For civil engineering applications that require very low differential settlements, a rigid slab is commonly employed.

The discrete element method was used to investigate the load transfer mechanisms within the embankment and to take into account the significant settlements of the compressible supporting soil, as well as the large deformations within the granular embankment and the soil–pile interaction mechanisms. Some of the advantages of the DEM compared, for example, to continuous methods, lie in the fact that it is possible to take account of complex mechanisms such as the punching of the granular embankment by the piles, the evolution of the mechanical behavior of the granular soil following the grains rearrangement and that this method integrates very large deformations.

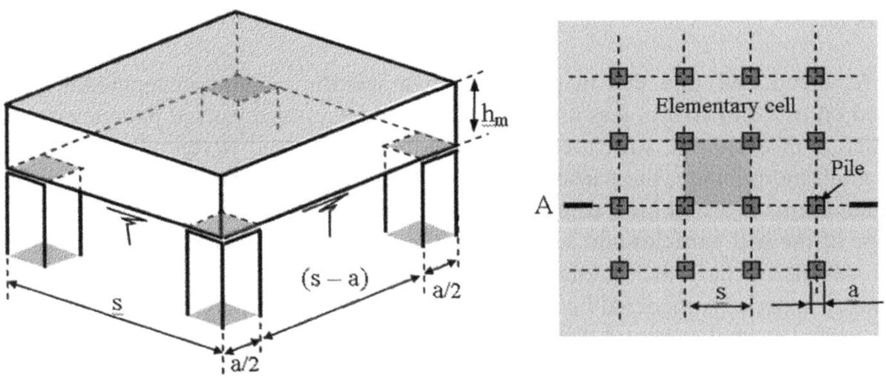

Figure 2.3. *Geometry of the network of piles [CHE 11]. For a color version of this figure, see www.iste.co.uk/grange/soil.zip*

Two applications will be proposed: the first concerns the behavior under monotonic loading of a granular layer over piles for the two following applications, under embankment and under building. The second concerns the study of the evolution of load transfer mechanisms under cyclic behavior of a granular layer above piles loaded by a rigid slab (wind turbine foundations, for example). The typical geometry of the proposed simulations is shown in Figure 2.3. For reasons of symmetry, a square mesh of rigid inclusions was considered. The pile caps are assumed as having a square section of side a. The distance between two piles is denoted s, and the covering rate of the network of piles (ratio of the surface of the pile cap to the area of an elementary cell of the network of piles) is defined as α.

The typical discrete numerical sample used is composed of (Figure 2.3):

– a granular embankment with a height h_m, made of a large amount of clusters. Depending on the cases, the soil elements of the numerical sample have been placed at different density states, according to the ERDF method, [CHA 05, SAL 09];

– for building applications, a rigid slab sited at the top surface of the embankment making it possible to apply monotone or cyclic loadings. This slab consists of two beds of spheres of the same diameter rigidly associated. Vertical forces applied at the upper part of the slab make it possible to obtain a rather uniform vertical stress;

– a compressible horizon whose action on the granular layer has been taken into account by vertical springs of stiffness K_c (expressed in MPa/m) making it possible to apply, at the base of the granular embankment, a vertical stress proportional to the settlement of the supporting soil (Winkler linear model). Thus, the relation between the settlement of the supporting soil Δh and the vertical stress σ_v applied to it is given by $\Delta h = \sigma_v/K_c$, where K_c is the coefficient of compressibility of the soil, which can be connected, for a subsoil layer of thickness H, to the oedometric modulus of the soil by the relation: $K_c = E_{oedo}/H$. The value of K_c is chosen so that centimeter displacements will be obtained during loading cycles. For example, a compressibility coefficient K_c of 0.75 MPa/m reflects the behavior of a compressible soil, 4 m thick, having an oedometric modulus of 3 MPa. Under these conditions, an unreinforced soil would undergo a settlement of 0.1 m under an overload of 75 kPa. It follows from this modeling that the soil is considered elastic and therefore the compressible horizon may, during the loading–unloading cycles, be compacted or disorganized freely, which would correspond to a configuration for which delayed settlements of the supporting soil are excluded;

– boundary conditions to take account of the symmetry conditions by way of frictionless vertical walls at the sides of the model. The piles are considered strongly rigid. As a result, the vertical displacements at the interface, between the granular material and the head of the piles, are fixed in the vertical direction. A Coulomb-type contact law was taken into account at the soil–pile interface to reproduce real friction mechanisms.

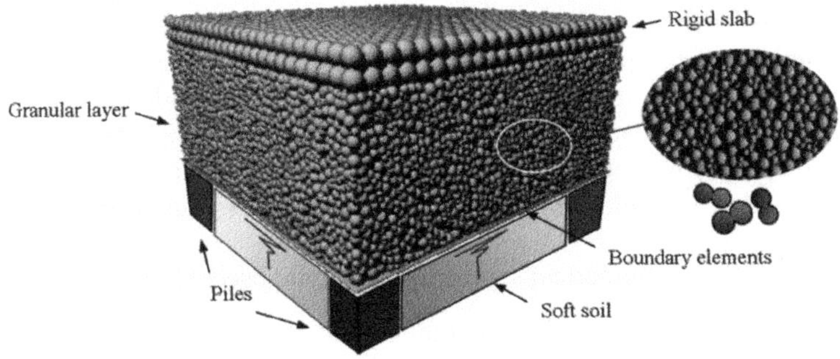

Figure 2.4. *Typical geometry of the numerical samples [CHE 11]*

Parameters investigated are the intensity of the load distributions on the piles and on the compressible soil, the surface settlements, the displacements of the granular particles and the load transfer mechanisms, which can be quantified by two specific parameters: the stress reduction rate, SRR, equation [2.6], defined as the ratio of the load F_p applied on one pile to the total vertical external loads (w + q) acting on an elementary cell of the network of piles, with w the slab and embankment weights and q the overload, and the ability G of the load transfer layer to postpone the overloads to the piles, equation [2.7], defined as the ratio of the incremental load ΔF_P acting on the piles to the overload applied q.

$$SRR = \frac{F_p}{w+q} \quad [2.6]$$

$$G = \frac{\Delta F_p}{q} \quad [2.7]$$

The results expected are the subsoil settlement, the displacements of the discrete particles, the network of contact forces within the granular embankment, the stress states at different points of the granular layer and the contact forces between the granular soil particles and the piles, in order to determine the efficiency of the pile network.

2.3.1. *Load transfer mechanisms within granular embankments over a network of piles*

This work was performed in the framework of the PhD of Bastien Chevalier [CHE 08], in collaboration with Gaël Combe, and of the Master's thesis of Quoc Anh Tran [TRA 15] in collaboration with Daniel Dias. Thanks to them.

Two applications will be presented successively: the first showing the arching effect that can be developed within a granular layer subjected to a uniform loading, and the second, the load transfer mechanisms obtained resulting from the use of a rigid loading slab [CHE 10, 11]. For these two applications, only a 2.5 m by 2.5 m elementary cell was considered. The pile horizontal cross-section is 0.375 m by 0.375 m. Thus, the covering rate represents 2.19%. The load transfer embankments are made up of an assembly of clumps, generated at the minimal porosity of 0.355 using the ERDF method [CHA 05, SAL 09].

Clumps are made of two imbricate spheres (of the same diameter d) whose centers have a distance of 0.95 d between them. The shape ratio of the greater to smaller particles is 4. Two thicknesses of the granular layer were considered: a load transfer layer of 0.5 m in height made up of 16,000 particles and the other 1.0 m in height made up of 32,000 particles. The micromechanical parameters and the corresponding macromechanical parameters are given in Table 2.1. A friction angle δ_i of 30° was considered at the interface between the granular elements and the piles. Four values of K_c were used to determine the influence of the subsoil compressibility on the load transfer mechanisms: K_c = 0.25, 0.5, 0.75 and 1.00 MPa/m. Several successive values of the uniform vertical overload q were applied: 12.8 kPa, 25.5 kPa, 46.8 kPa and 68 kPa. The expected load transfer mechanisms are due to the differential settlements produced by the contraction of the compressible soft soil.

Normal contact stiffness coefficient (N/m²)	K_{nij}	1.0×10^9
Ratio of tangential to normal contact stiffness (-)	K_{tij}/K_{nij}	0.75
Micromechanical friction coefficient (-)	$\mu = \tan \delta$	0.577
Young's modulus (MPa)	E	257
Poisson's ratio (-)	ν	0.08
Peak friction angle (°)	φ_p	44.8
Residual friction angle (°)	φ_r	30.1
Dry apparent density (kN/m³)	γ_d	17.7

Table 2.1. *Micro- and macromechanical parameters of the discrete model*

The numerical results obtained for SRR and G are presented in Figure 2.5 for the two studied granular embankment heights and for a subsoil compressibility coefficient K_c of 0.75 MPa/m. It can be seen in this figure that SRR increases with the total applied load ($q_t = w + q$) to reach a threshold value, dependent on the height of the granular embankment, that is much higher than the covering rate of 2.19%. At the same time, it is observed that the capacity G of the granular mattress to transfer efforts towards the piles is rather constant for given values of the granular embankment height (approximately 75% and 35% for embankment heights of 1.0 m and 0.5 m, respectively).

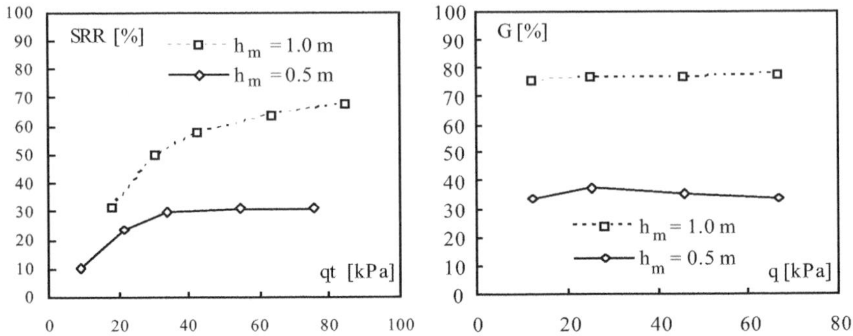

Figure 2.5. *Load transfer efficiencies expressed as the SRR and G ratios for K_c = 0.75 MPa/m and two values of h_m [CHE 11]*

The interaction mechanisms between the piles and the load transfers embankment are shown in Figure 2.6, by way of the intensities of the displacements of the particles of the granular material in a cross-section between two piles. In this figure, we clearly distinguish two areas: the first located above the piles where the movements are very low, the second above the compressible soil, where the displacements are rather similar to the surface settlement. As it can be seen, these two zones can be distinguished from each other by an inclined line, whose slope is directly correlated to the peak friction angle of the granular material.

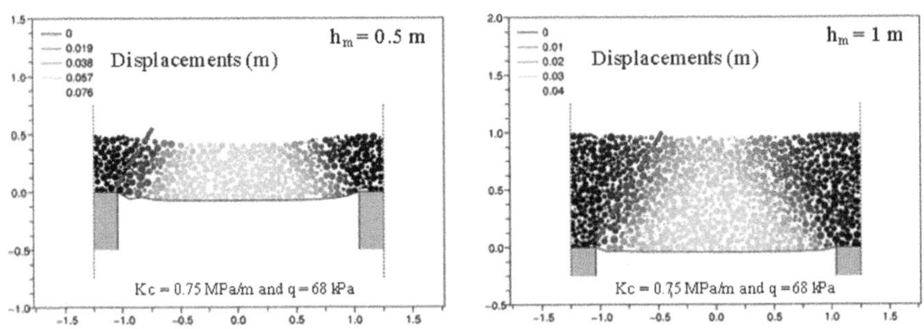

Figure 2.6. *Displacements of the granular particles in a cross-section sited between two piles for two values of h_m and Kc = 0.75 MPa [CHE 10]. For a color version of this figure, see www.iste.co.uk/grange/soil.zip*

Looking now into the case of the load transfer mechanisms when using a rigid slab, we present in Figures 2.7 and 2.8 the SRR reduction rates and the capacity G of the granular material to transmit the overloads to the piles. Due to the use of a rigid slab, only a 0.5 m-thick granular layer is tested. In these figures, it can be clearly seen that when the overload is applied on the granular embankment by a rigid slab, the SRR reduction ratio and the capacity G to retransmit the external forces to the piles are greater than in the previous case, where the overload is applied directly on the top surface of the granular embankment.

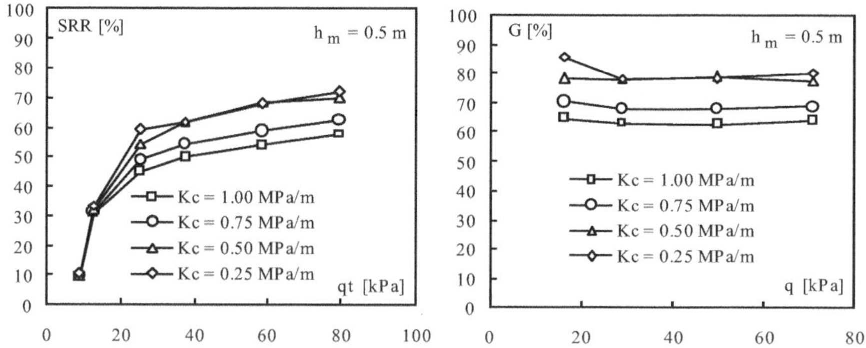

Figure 2.7. *Load transfer efficiencies when using a rigid slab expressed as the SRR and G ratios for different values of K_c and for h_m = 0. 5 m [CHE 11]*

If we compare the movements and displacements of the particles of the granular embankment for the two cases studied, with or without rigid slab (Figure 2.8), a significant difference can be seen in terms of mechanical behavior. For example, the most sheared soil zone, in the case of loading slab, is highly localized and is essentially limited to the soil cylinder above the head of the piles. This soil cylinder is highly compressed and the surface settlements obtained are strongly conditioned by its mechanical behavior.

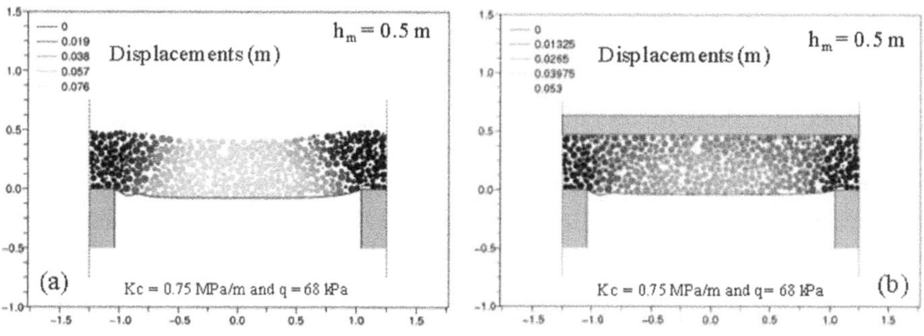

Figure 2.8. *Displacements of the granular particles in a cross-section sited between two piles for K_c = 0.75 MPa: (a) embankment and (b) under a rigid slab*

In order to better understand the influence of the soft soil compressibility on the efficiency, we compare, in Figure 2.9, the results obtained with and without a rigid slab for several values of the compressibility K_c. For very soft soils (K_c = 0.25 MPa/m and 0.5 MPa/m), and without a rigid slab, the efficiency increases with the total load and then decreases for greater values of the load.

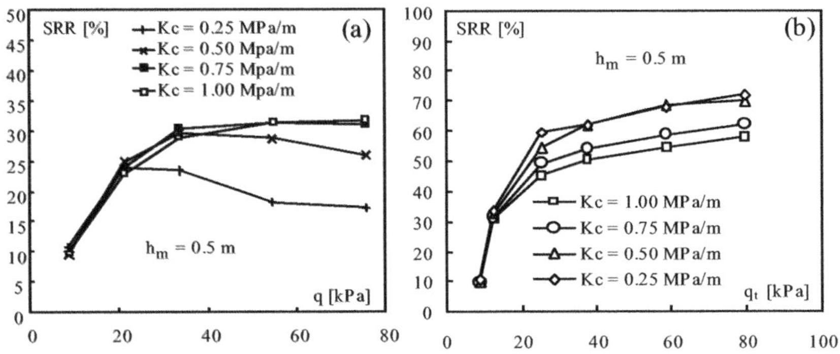

Figure 2.9. *Comparison of the values of SRR versus the total load applied q_t for H_m = 0.5 m and different values of the subsoil stiffness K_c: (a) embankment and (b) under a rigid slab*

For soft soils with a greater stiffness (K_c = 0.75 MPa/m and 1.0 MPa/m) and without a rigid slab, the load transfers increase regularly with the total load and then reach a threshold. For very soft soils (K_c = 0.25 MPa/m and 0.5 MPa/m) and without a rigid slab, the efficiency increases with the total load and then decreases for greater values of the load. Note that, in this case, the maximum efficiency increases with the increase of soft soil stiffness. In the case where the overload is transmitted to the granular material, by means of a rigid slab, the observations are different: the more the soil is compressible, the more transfers of loads to the piles are important. In fact, this is due to the ability of the stiffer subsoil to withstand the overload when a constant surface settlement is imposed by the rigid slab.

In order to compare the discrete element method and the finite difference method, particular care was taken to perform numerical simulations in an equivalent manner. Therefore, in the continuum approach, the cap yield model, accounting for friction hardening and softening, has been used to capture the behavior of the granular material. The main common parameters are: s = 3 m, a = 0.6 m, h_m = 0.75 m, 1.5 m, 2.25 m, 3 m and K_c varying between 0.05 to 1.0 MPa. Thus, the covering rate is equal to 4%. Only a quarter of the elementary mesh was considered for the two models, as presented in Figure 2.10. The DEM numerical samples were made of an assembly of clusters of various sizes (d_{max}/d_{min} = 4), composed of two overlapped spheres of diameter d with an angularity of 0.8 d; 16,000, 32,000, 48,000 and 64,000 clumps were used for the four considered values of h_m.

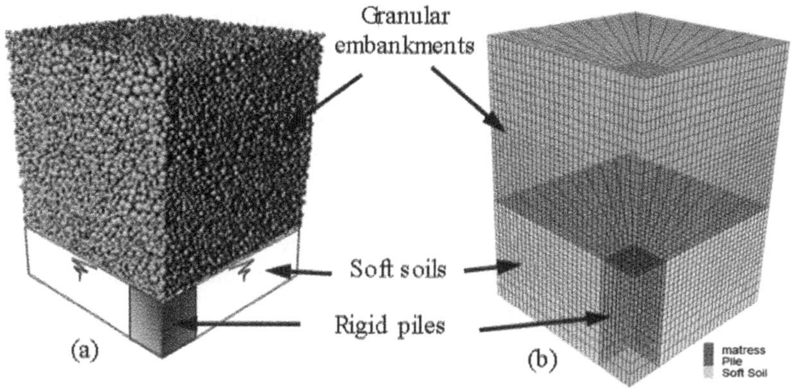

Figure 2.10. *Numerical samples used: (a) DEM and (b) FDM [TRA 19]. For a color version of this figure, see www.iste.co.uk/grange/soil.zip*

The DEM numerical sets of micromechanical parameters retained make it possible to restore the behavior of loose (L), medium (M) and dense (D) granular materials, as presented in Table 2.2. For the continuum model, the macromechanical parameters, achieved by comparison with the DEM triaxial tests, make it possible to reproduce rather similar behaviors of the three DEM numerical granular materials retained.

Numerical sample	**L**	**M**	**D**
Numerical porosity	0.41	0.38	0.34
K_{nij} **(N/m²)**	1.0 x 10^7	1.0 x 10^7	1.0 x 10^7
K_{tij}/K_{nij}	1	1	1
$\mu = \tan\delta$	0.84	0.84	0.84
ϕ_p **(°)**	34	40	46
ϕ_r **(°)**	26	26	26
γ_d **(kN/m³)**	1.47	1.55	1.65

Table 2.2. *Micro- and macromechanical parameters of the discrete model*

First comparisons between the two numerical models were made in Figure 2.11a, considering the efficacy SRR of the load transfers versus the soft soil stiffness (K_c).

For each case tested, a maximal value of the efficacy can be found, for which the intensity is a function of the subsoil stiffness. For high values of the subsoil stiffness, the load resulting from the weight of the granular embankment is transferred mainly to the subsoil, strong enough to support the load applied. When the stiffness of the subsoil decreases, the load transmitted to the piles increases to a maximum value, characteristic of the maximum load transfer. For very low values of the subsoil stiffness, the efficacy reduces dramatically due to the punching of the granular embankment by the pile. In this case, some discrepancies between discrete and continuum models were observed, due to the large displacement at the soil–pile interface during the punching of the granular material. In this case, the DEM model seems to be more relevant to take account of the large displacements of the embankment, especially in the vicinity of the corner of the pile cap. The maximal efficacy as a function of the density of the granular material, characterized by its peak friction angle, is presented in Figure 2.11b.

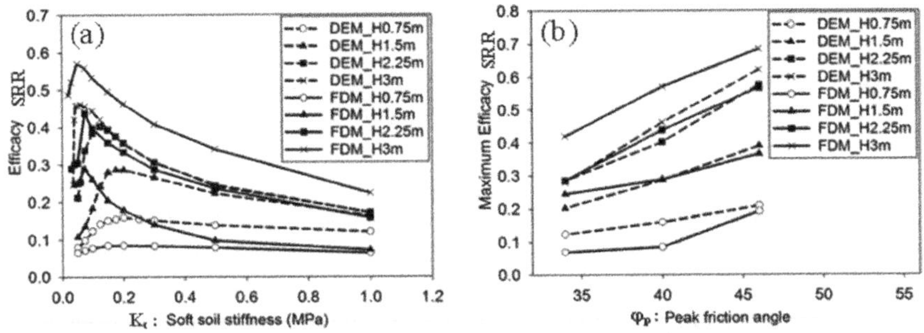

Figure 2.11. *Efficacy versus subsoil stiffness for the material density M and maximal efficacy obtained for all the numerical simulations performed [TRA 19]*

As can be seen in this figure, the comparison between DEM and FDM shows similar trends and values for the maximal efficacy, especially in the case of dense material, regardless of the embankment height. As can be expected, for a fixed value of K_c, the maximal efficacy increases with the density and the thickness of the embankment. Load transfer mechanisms can also be seen when looking at the curve representing the efficacies SRR versus the shear rate (Figure 2.12), defined as the ratio between the maximum vertical displacement of the granular embankment and the embankment height h_m. It can be seen that the efficiency increases with the shearing ratio until the maximal efficacy value was obtained. When large shear mechanisms are obtained, the efficacy decreases as a function of the density of the granular embankment: a dense material leads to a greater decrease than a loose material. These trends can be observed for both the FDM and DEM, despite the disparity obtained between the two numerical models.

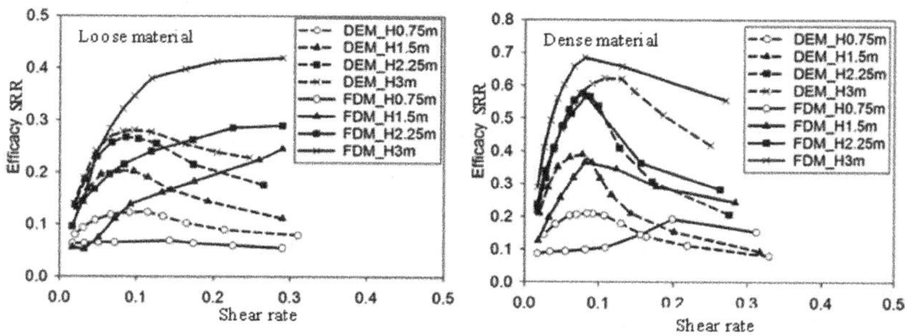

Figure 2.12. *Efficacy versus shear rate obtained for all the numerical simulations performed for the material densities L and D [TRA 19]*

The considerable reduction of the load transfer efficacy, observed in the discrete model when dealing with great values of the shearing strain, cannot be reproduced by the continuum model, especially for loose material. This can be attributed to the ability of the DEM model to take account of the change in the mechanical behavior of soil when large shearing and expansion of the granular soil occur. Moreover, the corner of the pile cap represents a singularity for the FDM that prevents sliding of the soil near the pile and introduces force concentration, which is not observed with DEM.

2.3.2. *Load transfer mechanisms within granular embankments over a network of piles under cyclic loadings*

This work was performed in the framework of the engineer internship of Agathe Furet [FUR 13] in collaboration with Stéphane Grange. Thanks to them.

The potentiality of discrete element models to account for load transfer mechanisms has already been tested and validated under monotonic loading. The possible application of the numerical model to cyclic solicitations results from the fact that the discrete element method explicitly takes into account the mechanisms that develop at the micromechanical level, such as the irreversible displacements between particles and the granular assembly rearrangement at each cycle, without the need to implement complex interaction laws. For this reason, a simple Coulomb-type frictional model is used in this study. To avoid dynamic phenomena, the loading speeds (in force or in displacement), which can lead to rather long simulation durations, must be sufficiently weak, especially when a large number of loading cycles occur.

The work presented here [FUR 13] focuses on the evolution of the load transfer mechanisms within a granular embankment loaded by a rigid slab and subjected to a large variety of cycles. Note that the interaction contact law used leads – when an inversion of the relative displacement between the grains is initiated – to a gradual decrease in tangential contact forces, until a complete change of orientation is obtained when the relative displacements between grains increase significantly. Moreover, when the maximum shear is reached and the normal contact forces between the soil particles decrease, which will be the case during loading–unloading cycles, there is logically a reduction in shear forces. Under these conditions, it is clear that loading–unloading cycles will generate a large variation in the normal and tangential contact forces between the grains and modify the load transfer mechanisms.

For this study, the compressible soil is elastic, and because of this, during unloading cycles, can be discharged, which will generate changes in the structure of the granular assembly and thus disrupt the mechanisms of load transfer. The applications targeted by this type of modeling correspond to cyclic loadings cases, for which delayed settlements are excluded. In the present study, only one subsoil stiffness and several loading cycles of different amplitudes were tested.

The geometry of the network of piles refers to a square mesh of rigid inclusions for which the distance between two adjacent piles is s = 2.5 m. The piles have a square section of side a = 0.416 m. The covering rate of the network of piles is 2.77%. The embankment height is 0.5 m. For reasons of symmetry, only one elementary cell of 2.5 m by 2.5 m was considered for modeling. The discrete numerical sample is composed of 16,000 clusters made of two imbricate spheres of the same diameter d spacing of 0.95 d, as defined previously. A compressibility coefficient K_c of 0.75 MPa/m was retained for the subsoil stiffness, and a friction angle of 30° was taken into account at the soil–pile interface. The micromechanical and macromechanical parameters are similar to those defined previously (Table 2.1).

The numerical simulations were conducted according to the following processes:

– application of gravity to the particles of the granular material and of the loading slab with the assumption of zero vertical displacement at the base of the embankment. The density of the spheres is such that the initial vertical stress applied by the slab on the granular material is 7 kPa;

– slow gradual displacement of the particles of the granular material due to gravity until their stabilization without dynamic effect on the compressible soil;

– constant-amplitude cycles, from 0 to 60 kPa (load case A) and from 60 to 120 kPa (load case B), are imposed in successive increments on the slab. The main objective of these simulations is to evaluate the influence of the loading rate of the embankment on the evolution of the load transfer mechanisms. About 150 loading cycles were performed. Several loading speeds have been tested to verify that within the selected speed range, the dynamic effects are minimized.

The results obtained in terms of efficiency of the load transfer mechanisms are shown in Figures 2.13, 2.14 and 2.15. q_t is the total stress applied comprising the weight of the granular material, the weight of the loading slab and the overload, and d_s is the vertical displacement of the slab during the loading–unloading cycles. Figure 2.13 shows the evolution of the vertical displacements of the slab during the loading–unloading cycles. It can be seen that the displacements of the slab are rather small but that they change over time during the loading cycles, and this in a similar way for the two load cases studied. This increase in displacement of the slab during loading cycles reflects a loss of efficiency of the load transfer mechanisms.

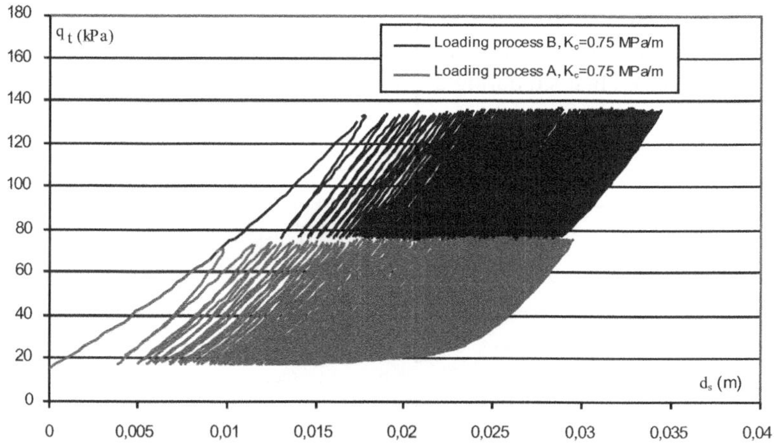

Figure 2.13. *Evolution of the vertical displacements of the slab (d_S) as a function of the total load applied for the two selected loading processes (K_C = 0.75 MPa/m). For a color version of this figure, see www.iste.co.uk/grange/soil.zip*

The load transfer rates SRR are presented in the following figures, according to the displacement of the loading slab (Figure 2.14) and to the number of cycles (Figure 2.15). In Figure 2.14, we see that 50% of the loads resulting from the weight of the embankment and of the loading slab are redirected to the piles before the cyclic loading process. Subsequently, the influence of the loading slab on the load transfer is clearly highlighted since the percentage of the total load redirected to the piles increases progressively during the first loading, until reaching a maximum value of 85% for the loading process used.

At the first loading cycle, it can be seen that the efficiency of the load transfer system (granular embankment and loading slab) decreases during the unloading phase as a result of the change in structure of the granular material and of the network of contact forces following the elastic unloading of the compressible soil. Unloading the granular embankment implies the reduction or cancellation of the tangential contact forces between the particles, following rearrangement and relative displacements of the grains during unloading.

For loading process B (cycles from 60 to 120 kPa), this decrease in efficiency remains low (approximately 8%). For loading process A (cycles from 0 to 60 kPa), this decrease is much larger considering that the granular layer is at each cycle completely discharged. After the first loading cycle, the efficiency rate SRR obtained is, in this case, about 40%. This value is to be compared with the value of 50%, corresponding to the efficiency of the load transfer system before setting up the loading cycles. The load transfer mechanisms generated within the granular material have therefore been altered in the first cycle and will continue to degrade over all loading cycles, given that the efficiency of the load transfer system never reaches the level of performance obtained during the phase related to the first loading.

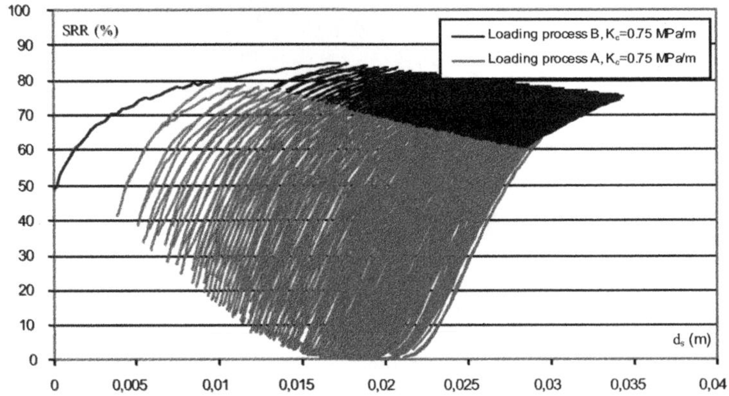

Figure 2.14. *Evolution of the efficiency of the load transfer system SRR as a function of the displacements of the slab d_s for the two loading processes. For a color version of this figure, see www.iste.co.uk/grange/soil.zip*

In Figure 2.15, we see (loading processes A and B) a gradual decrease in the maximum efficiency of the load transfer system over the first 150 cycles studied. This loss is about 10% over the entire loading–unloading cycles, which remains acceptable given the high efficiency rates obtained. In the unloading phase, the contrast in behavior between the two loading processes studied is very important. This is related to the fact that in loading process B, the granular embankment is not completely discharged, whereas in loading process A, it is completely unloaded.

The almost complete cancellation of the efficiency after unloading in case A means that the granular embankment is highly unstructured during the loading–unloading cycles and that the load transfer mechanisms that have been set up in the embankment, during the first phase of the numerical process (application of the gravity), were completely annihilated. On the other hand, after reloading the embankment with an overload of 60 kPa, it can be seen that the efficiencies obtained in loading process A are very close to those obtained after the unloading cycles in the case of loading process B (overload of 60 kPa).

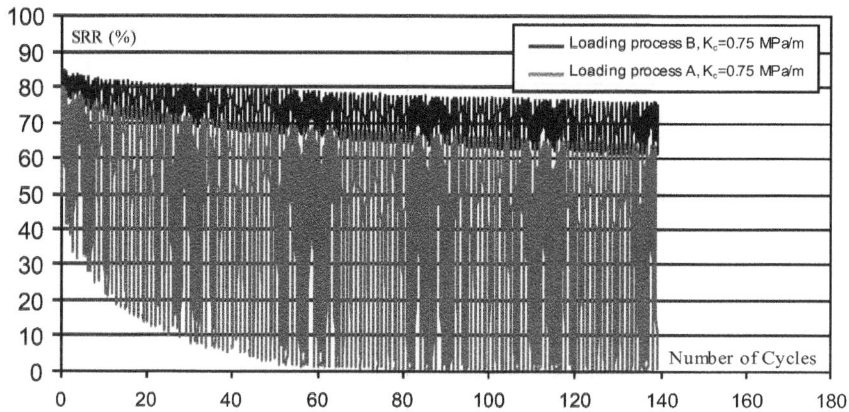

Figure 2.15. *Evolution of the efficiency of the load transfer system SRR as a function of the number of loading cycles for the two loading processes. For a color version of this figure, see www.iste.co.uk/grange/soil.zip*

To better distinguish the role of the deformation of the granular embankment on the load transfer mechanisms, we have shown, in Figure 2.16, the efficiency curves G of the load transfer system as a function of the displacement of the slab at each loading cycle. Recall that G represents the capacity of the granular material to postpone the overloads to the piles (and not the total load). It is defined as the ratio of the load increment ΔF_P acting on the piles to the overload increment applied Δq and therefore ignores the load transfer mechanisms that occur during the set up of the embankment and the loading slab.

It can be seen from this figure that for an equivalent deformation of the granular material (d_s unchanged), the capacity of the granular material to redirect the forces towards the piles is identical to and independent of the two loading processes studied. For a given deformation state, the capacity of the granular layer to mobilize the load transfers is therefore unchanged. It will also be noted that the loading–unloading cycles gradually disorganize the granular material, resulting in a loss of efficiency G, which varies from 90% to 80% for all the cycles performed. It is concluded that it is the changes in the granular structure and the rearrangement of the grains that condition the load transfer mechanisms.

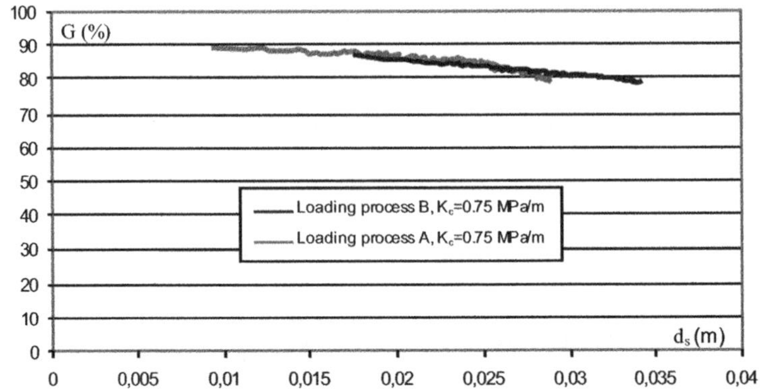

Figure 2.16. *Efficiency of the load transfer system G to redirect overloads to the piles according to slab displacements d_s for the two selected loading processes. For a color version of this figure, see www.iste.co.uk/grange/soil.zip*

The numerical results show that, for the value of the subsoil stiffness tested and for the levels of the amplitudes of the loading cycles carried out, the efficiency of the granular embankment gradually decreases during the loading–unloading cycles following a disorganization of its structure and modifications of the tangential contact forces between particles. The efficiency values obtained at each loading cycle nevertheless remain significant, even after 150 cycles. When the loading cycles lead to the complete suppression of the overload (loading process A), it is demonstrated that after a series of loading–unloading cycles, the load transfer mechanisms that have developed in the load transfer system (following the set up of the granular material and the loading slab) are completely annihilated, which means that after a certain number of cycles, the total weight of the granular material and the slab is supported by the compressible soil.

2.4. Application of the DEM to geotechnical structures in interaction with flexible and deformable reinforcement – comparison with experiment results

Due to their ease of implementation and their very competitive cost, geosynthetic reinforcements are widely used for the reinforcement of geostructures. Geosynthetics are most often in the form of a plane sheet made of woven fibers, knitted or bonded together and which can mobilize, by their tensioning, tensile forces that cannot be supported by the soil. To model these deformable reinforcements, thin triangular flat elements, [GIR 97, VIL 98], defined by three nodes, were implemented in the DEM computation code, [LEH 06, VIL 09]. To ensure the continuity and progressiveness of the frictional forces from one element to another during stretching of the sheet, cylinders and spheres have been positioned, respectively, on the edges and at the nodes of each triangular element (Figure 2.17).

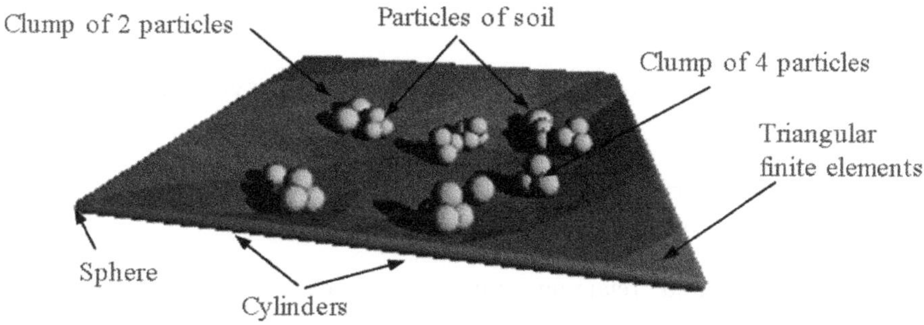

Figure 2.17. *Finite elements used to ensure the geometric continuity of the geosynthetic sheet*

The behavior of the geosynthetic elements is governed by a stiffness matrix [VIL 98] that takes into account each orientation and behavior of the fibers. The basic assumptions of the formulation of these elements are:

– each element consists of a set of fibers with various orientations, initially forming a plan;

– there is no slipping between the fibers (presence of connection points between wires);

– the tensile forces acting in each fiber are oriented in the direction of this fiber after deformation (large displacements).

The numerical parameters required for modeling the geosynthetic layer are: the tensile stiffness of the geosynthetic in each fiber direction and the interface parameters (k_{ni}, k_{si} and δ_ι), as previously defined, between a soil particle and a plane. The behavior of the geosynthetic elements is managed, as for conventional discrete elements, by the alternation of Newton's second law of motion applied to the nodes of the elements, and the determination of the interaction forces. In the case of a continuous geosynthetic sheet, the interaction forces that need to be taken into account are those acting between the sheet elements (following their tensioning) and those generated by the contact with the soil particles.

Two applications to soil structures reinforced by geotextiles will be presented, for which the interaction mechanisms between the soil and the reinforcement are essential to their survival. The first refers to retaining walls whose facing is made of geotechnical tubes filled with a granular material. The second concerns road and rail embankments, reinforced by geosynthetic sheets subjected to localized sinkholes.

2.4.1. *Numerical and experimental behavior of geosynthetic tubes filled with granular material*

This work was performed in the framework of the PhD of Joanna Gorniak [GOR 13]. Thanks to her.

The need to build retaining structures in areas with weak mechanical characteristics led to the testing and construction of reinforced geostructures, made of lightweight aggregates, encapsulated within geosynthetic tubes. These structures are often used in geotechnical engineering mainly to reduce the bearing pressure acting on the subgrades, while providing a high stability and significantly reducing the settlements [WAT 04]. A quasi-vertical slope of the front of the reinforced structures can be realized using stacked geosynthetic tubes, filled with lightweight aggregates (Figure 2.18).

Figure 2.18. *Construction phase of a retaining wall with a facing made of geosynthetic tubes filled with granular material [GOR 15]. For a color version of this figure, see www.iste.co.uk/grange/soil.zip*

These tubes are filled quickly, at a constant speed, using a specific process developed in the framework of the TeMaSi research program, enabling a homogeneous and uniform shape of the tube cross-section. The compaction process needs only three or four passes of a compactor and provides a regular and constant height of the tube. During these tests, it has been observed that the filling rate of the tube is also dependent on the pressure from the blowing truck and the developed filling device.

In situ experiments were carried out to analyze the behavior of these structures under different types of loading. In particular, the geosynthetic tubes, used for the facing of the retained structures, were the subject of specific instrumented experiments of loading and unloading (Figure 2.19). The main concept of this technique is that the geosynthetic tube provides a confining pressure to the encapsulated granular clay material consecutively to the application of a vertical loading, allowing the mobilization of tensile forces in the geosynthetic. To complete this study, discrete numerical models (Figure 2.20) were used to better understand the contribution of each of the components of the granular filled tubes, and to allow the establishment of a design method.

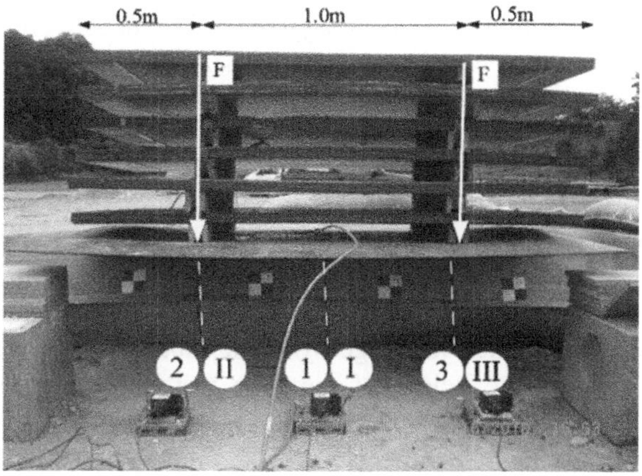

Figure 2.19. *Loading test performed on a geosynthetic tube filled with granular material [GOR 16]. For a color version of this figure, see www.iste.co.uk/grange/soil.zip*

The experimental loading tests are performed on a 6 m-long geosynthetic tube filled with lightweight granular materials. Several types of wrap-knitted geotextiles, produced from different polymers, were used (referred to as PP-PVA, PVA, PP-1, PP-2 and PES) with rather similar mechanical properties, i.e. an initial stiffness in the radius direction of 59 kN/m and an initial stiffness in the longitudinal direction of 43 kN/m.

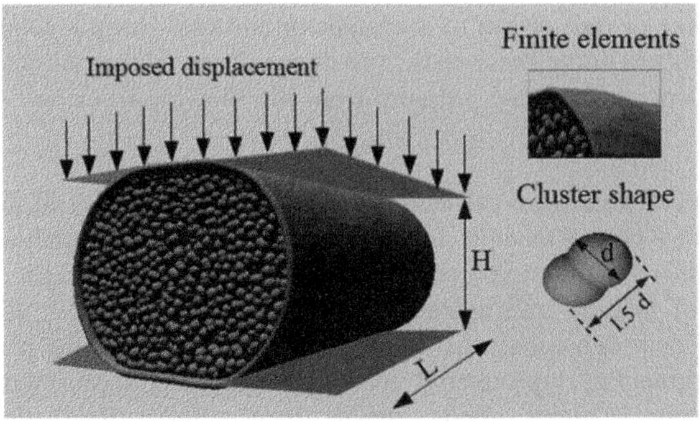

Figure 2.20. *Geometry of the numerical DEM sample [GOR 13a]. For a color version of this figure, see www.iste.co.uk/grange/soil.zip*

The aggregates are expanded clay aggregates, produced on the principle of expansion of soft clay (LWA), showing a grain size distribution of 10–20 mm and an internal friction angle between 35° and 38° [GOR 16]. Two main filling rates (85% and 100%) were obtained depending on the materials and filling device used. The filling rate of 100% was considered for the first group of tubes, constructed of PP-PVA, PVA and PP-1 for a bulk density of 540–560 kg/m^3, while the second group, represented by tubes PP-2 and PES, has a bulk density of 490–498 kg/m^3. A steel plate, positioned in the central part of the tube, makes it possible to apply a vertical load, which was measured by load cell transducers. The total compressive load applied to the geosynthetic tube equals 31 kN/m. The cross-sectional geometry and the height of the tube were obtained from 3D scanner measurements after each loading step.

The numerical loading is performed on 1.0-meter-long tubes with initial radius of $R_0 = 0.3$ m. Smooth vertical, rigid plates at both ends of the tube ensure the plane strain boundary condition. To reproduce the macroscopic behavior of expanded clay aggregates, 12,000 clusters made of two overlapping spheres of diameter d are used. The distance between the centers of the two spheres of one cluster is equal to 0.5 d. A ratio of 2 between the maximal and minimal diameters was used. The particles of the granular material are generated at different porosities η using the expansion radius and decrease of friction process (ERDF), in order to simulate different rates of filling of the granular material included within the geosynthetic tubes. Values of the micromechanical parameters and corresponding macromechanical parameters are given in Table 2.3 for the five values of the tested porosities. E is the initial Young's modulus, ϕ_p is the friction angle at the peak, ϕ_r the residual friction angle at the threshold, γ_{peak} is the angle of dilatancy defined between 2% and 3% of vertical strain and α is the residual volumetric deformation obtained at the threshold. A total of 768 thin three-node triangular finite elements, forming a continuous fabric, are used to describe the tensile and membrane behavior of the fabric. The interaction between the discrete particles and the elements of the geosynthetic tube is taken into account by a friction contact law, characterized by: E_0 = 5 MN/m^3 and a friction parameter φ_i equal to 0° or 30° depending of the case studied. The stiffness of the geotextile tube is, according to those of the experiment, 59 kN/m in the radial direction and 43 kN/m in the longitudinal direction. Loading forces are imposed by the regular vertical displacements of two rigid horizontal plates that interact with the geosynthetic fabric on the top and at the bottom of the tube.

<table>
<tr><th></th><th colspan="7">Micromechanical parameters of the numerical sample</th></tr>
<tr><th></th><th>η
(-)</th><th colspan="2">D_r
(%)</th><th colspan="2">K_{nij}
(N/m²)</th><th>K_{sij}
(N/m²)</th><th>$\mu = tg\ \delta$
(-)</th></tr>
<tr><td>Soil-1</td><td>0.34</td><td colspan="2">78.2</td><td colspan="2" rowspan="5">1.0 10^8</td><td rowspan="5">1.0 10^8</td><td rowspan="5">1</td></tr>
<tr><td>Soil-2</td><td>0.36</td><td colspan="2">60.1</td></tr>
<tr><td>Soil-3</td><td>0.38</td><td colspan="2">43.5</td></tr>
<tr><td>Soil-4</td><td>0.40</td><td colspan="2">26.1</td></tr>
<tr><td>Soil-5</td><td>0.42</td><td colspan="2">8.7</td></tr>
<tr><th></th><th colspan="7">Macromechanical parameters of the numerical granular assembly</th></tr>
<tr><th></th><th>D_r
(%)</th><th>E
(MPa)</th><th>ϕ_p
(°)</th><th>ϕ_r
(°)</th><th>γ_{peak}
(°)</th><th>α
(%)</th><th>ν
(-)</th></tr>
<tr><td>Soil-1</td><td>78.2</td><td>29</td><td>46</td><td rowspan="5">26</td><td>23</td><td>15</td><td>0.081</td></tr>
<tr><td>Soil-2</td><td>60.1</td><td>24</td><td>42</td><td>19</td><td>11</td><td>0.091</td></tr>
<tr><td>Soil-3</td><td>43.5</td><td>20</td><td>38</td><td>16</td><td>7.2</td><td>0.113</td></tr>
<tr><td>Soil-4</td><td>26.1</td><td>15</td><td>34</td><td>15</td><td>3.2</td><td>0.139</td></tr>
<tr><td>Soil-5</td><td>8.7</td><td>11</td><td>28</td><td>8</td><td>-0.5</td><td>0.162</td></tr>
</table>

Table 2.3. *Micro- and macromechanical parameters of the numerical granular materials used*

To compare the experimental results to the numerical ones, a sensitivity study was performed, based on the influence of the filling rate of the granular material and its density and thus, its ability to expand or contract during shearing. This was motivated by the fact that the experimental filling rates were not well known, even if the experimental loading tests can be distinguished into two groups: tubes filled at 100% maximum rate and tubes filled at 85%. As a consequence, five numerical materials (soil-1 to soil-5) set up at different porosities were used as granular filling material. First, the own weight of the granular material was neglected regardless of the high intensity of the overload applied and the friction parameters φ_i between the loading plate and the geosynthetic tube, and between the soil and the geosynthetic, were set to zero for the sake of simplicity and due to the lack of relevant measurement; the influences of these parameters will be investigated later.

The experimental data obtained after the maximal loading (F_v = 34.5 kN/m) are compared in Figure 2.21 to the numerical results (loading curves defined as the height of the geosynthetic tube versus the compressive force). As can be seen, a relatively good correlation between the experimental and numerical results can be found. A clear distinction between the two degrees of filling can be made. It can be seen that the geosynthetic tube filled with the numerical granular material set up with a high density (soil-1) is significantly more rigid than the geosynthetic tube constituted by the granular material (soil-5), which has a high porosity. It can be also seen that the experimental results obtained with the first group of tubes made of PP-PVA, PVA and PP-1 for a bulk density of 540–560 kg/m^3 are in accordance with the numerical results obtained with dense numerical materials (soil-2 to soil-4), while the experimental results obtained with the second group of tubes, represented by tubes PP-2 and PES, for a bulk density of 490–498 kg/m^3, are logically well restored by the loose numerical material (soil-5).

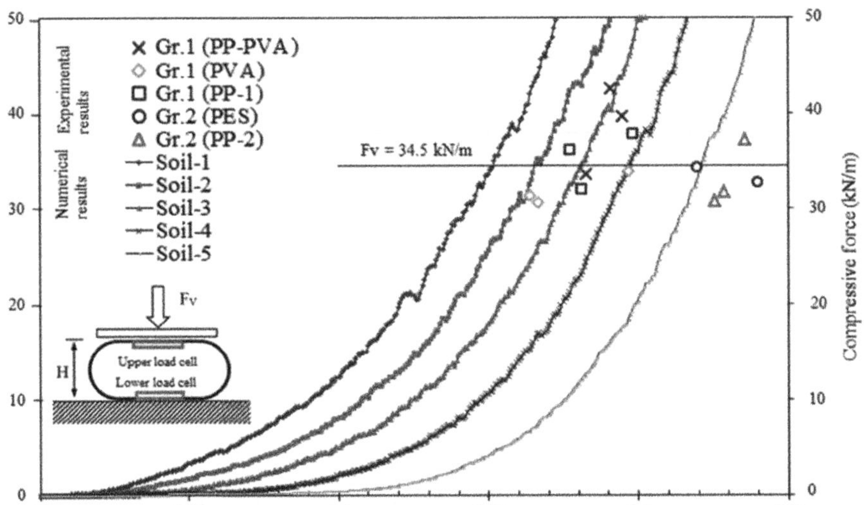

Figure 2.21. *Comparison between the experiment results obtained for F_V = 34.5 kN/m and the numerical force–displacement curves (soil-1 to soil-5) [GOR 16]. For a color version of this figure, see www.iste.co.uk/grange/soil.zip*

In order to better understand the role of the geosynthetic–soil interaction, we perform new loading tests following a protocol quite similar to the one previously used, which consists of a vertical compression up to 30% of axial deformation, decompression until a zero-compressive force and reloading up to 50% of the vertical deformation.

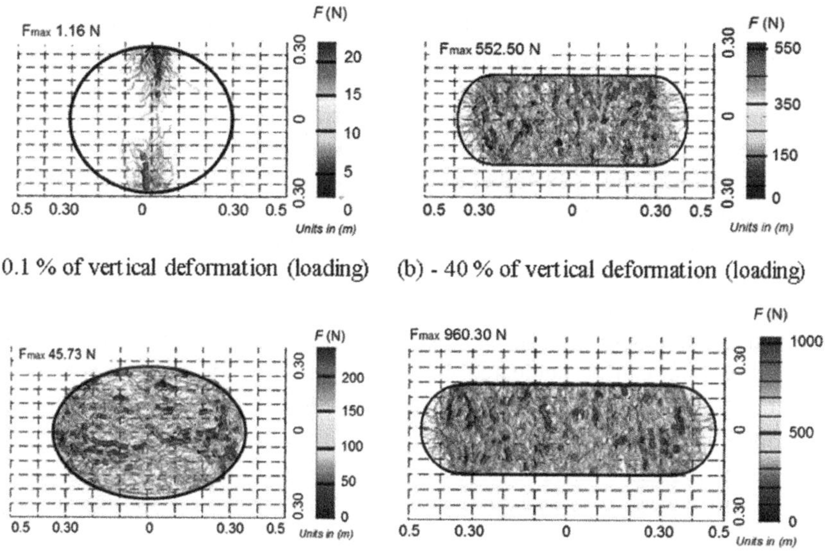

(a) - 0.1 % of vertical deformation (loading) (b) - 40 % of vertical deformation (loading)

(c) - Unloading to F_v=0 kN/m (unloading) (d) - 50% of vertical deformation (reloading)

Figure 2.22. *Network of contact forces between the granular particles at various stages of the loading–unloading numerical process [GOR 16]. For a color version of this figure, see www.iste.co.uk/grange/soil.zip*

The networks of contact forces obtained at different stages of the numerical loading–unloading–reloading process are shown in Figure 2.22. At the very beginning of the loading test, it can be seen that the interaction forces between the soil particles are highly concentrated in the vicinity of the contact points between the loading plates and the geosynthetic tube. This zone expands rapidly following the increase in loading; the central rectangular part of the filling tube being subjected to rather uniform vertical and horizontal stresses. The stress tensor in this area can be computed using equation [2.4]. From that, it can be seen that the horizontal and vertical stresses are the minor and minimum stresses, respectively. The ratio of the horizontal to vertical stresses is equal to 0.32 at 40% of the vertical deformation that corresponds to the active pressure coefficient of the numerical granular material. After unloading, this ratio equals 2.75, which refers to the passive pressure coefficient of the numerical soil. The ability of the geosynthetic tube to withstand vertical forces is obviously related to the rigidity of the geosynthetic reinforcement, which by confinement effect makes it possible to maintain a high lateral pressure. During unloading, the horizontal stresses remain, thanks to the beneficial action of the geosynthetic, whereas the vertical stresses logically vanished. The reloading of the tube leads to a state of stress similar to that obtained during the first loading. We will already underline the ability of the numerical model to account for the very large deformations of the geosynthetic tube and of the granular particles assembly.

Additional numerical simulations were carried out in order to better understand the influence of the friction effects (not taken into account previously) at the interface between the geosynthetic fabric and the filling soil, and between the loading plate and the supporting soil, using a friction coefficient set at 30°. As can be seen in Figure 2.23 (a – without friction, b – with friction between the fabric and the filling soil, c – with friction between the fabric and the loading plate and the foundation soil and d – with friction between the fabric and the soil, the loading plate and the foundation soil), the friction between the geosynthetic fabric and the granular material can be neglected without any significant influence on the results. The influence of the friction between the geosynthetic fabric and the supporting soil on the loading plate is very important.

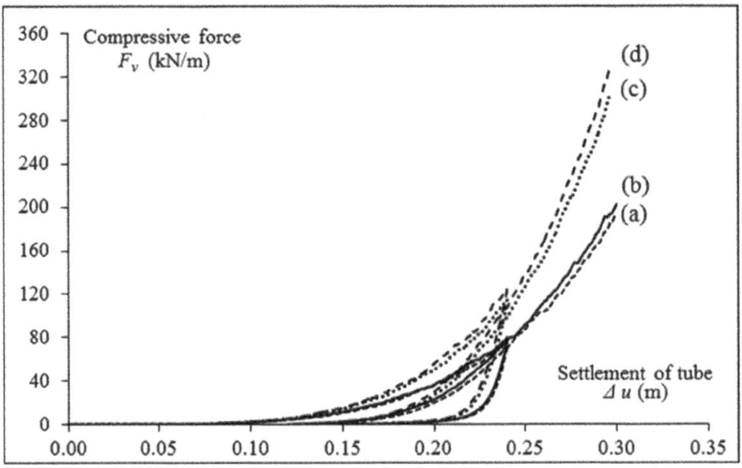

Figure 2.23. *Comparison of the numerical loading curves assuming different friction parameters between supporting soil, loading plate, geosynthetic fabric and granular material [GOR 13a]*

The influence of the friction on the distribution of the circumferential tensile force in the geosynthetic fabric, at 50% of the axial deformation, is presented in Figure 2.24. It can be seen that without friction between the loading plate, the supporting soil and the geosynthetic fabric (cases a and b), the distribution of tensile stresses is rather constant along the circumference of the tube. The presence of friction between the geosynthetic tube, the loading plate and the supporting soil (cases c and d) has both a significant effect on the loading capacity of the geosynthetic tube and on the tensile force distribution along the circumference of the tube. As can be seen, the friction forces have a beneficial effect on the loading capacity of the geosynthetic tube when subjected to vertical loading.

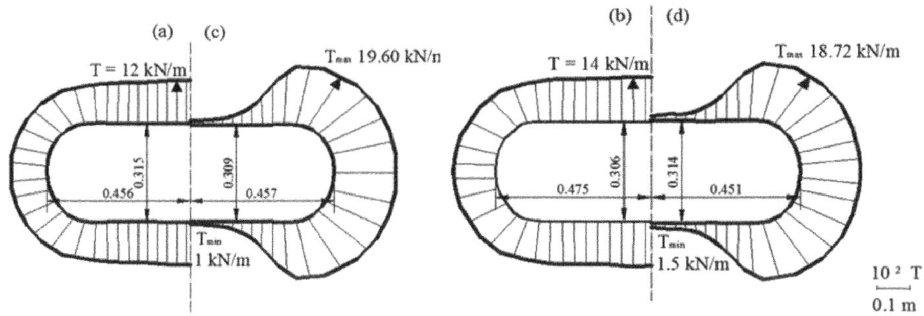

Figure 2.24. *Comparison of the numerical tensile forces in the geosynthetic fabric assuming different friction parameters between supporting soil, loading plate, geosynthetic fabric and granular material [GOR 13a]*

An analytical design method was developed on the basis of the true scale experiments and of the numerical results, in order to obtain a practical tool that can be used in engineering. This design method will be used for comparison and validation of the numerical model using simplifying assumptions. Particular attention was paid to estimate the loading capacity of the tube and to the values of the average tensile forces in the geosynthetic. The main assumptions used for the establishment of the loading force–displacement relation are:

– the geosynthetic tube is a three-dimensional infinite long elastic membrane of initial radius R_0 subjected to vertical external forces;

– the geosynthetic fabric is considered as a sufficiently thin membrane with no compressive and bending stiffness. The tensile behavior of the geosynthetic is governed by an elastic law ($T = J\ \varepsilon$);

– the filling material is a frictional granular soil with a friction angle φ and an expansive coefficient at large deformations denoted α. Taking into consideration a dilatant soil leads to the assumption that the increment of tube's volume is proportional to the initial volume. During the loading and unloading process, the granular soil is assumed to be at the limit state of its stability;

– the supporting soil and the loading plate are modeled as frictionless and rigid wall;

– the unit weight of the structure is neglected in view of the intensity of the applied vertical load, as well as the friction forces between the filling soil and the geosynthetic.

From these assumptions, and by studying the static equilibrium of any part of the loaded tube, it can be demonstrated [GOR 15] that:

$$T = \frac{J\left(4R_0^2(1+\alpha) + H^2 - 4R_0 H\right)}{4HR_0} \quad [2.8]$$

$$F_v = \frac{\pi J\left(4R_0^2(1+\alpha) - H^2\right)\left(4R_0^2(1+\alpha) + H^2 - 4R_0 H\right)}{8H^3 R_0 K_a} \quad [2.9]$$

where T is the average tensile force in the geosynthetic, equation [2.8], F_v is the load-bearing capacity of the filled tube, equation [2.9], H is the height of the deformed tube and K_a is the coefficient of the active soil pressure.

The comparison between the results of the numerical model and of the analytical formulation is represented as load–settlement curves (Figure 2.25). Similar assumptions were used in both cases, so that the weight of the granular material was neglected and the friction between the plate, the supporting soil, the geosynthetic fabric and the granular material was set to zero.

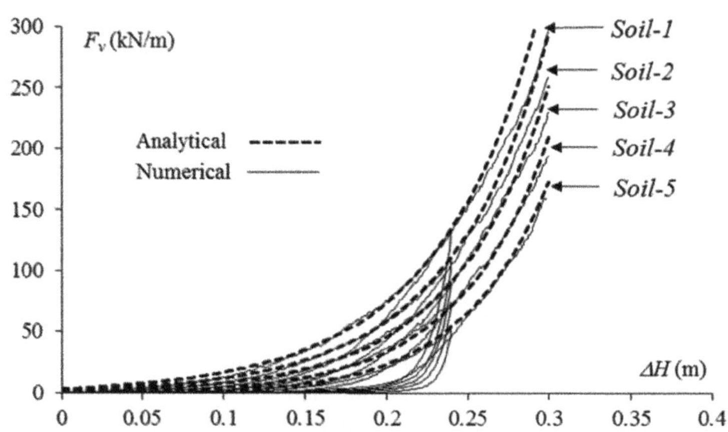

Figure 2.25. *Comparison between the numerical and analytical curves in terms of compressive force versus settlement [GOR 16]*

The five numerical granular materials (soil-1 to soil-5) were considered in order to test the ability of the numerical model to reproduce the loading behavior of the granular material filling tube, using different soil behaviors (expansive or contractive soils). Due to the large deformation of the geosynthetic tube, the comparison was established using the characteristics of the numerical granular soils obtained under large displacements ($\varphi = \varphi_r = 26°$). The expansion coefficients α used for the analytical design method are those presented in Table 2.3.

From the obtained results, it was shown that the loading capacity of the geosynthetic tube is a function of the expansive coefficients of the granular material. Except for very large values of the expansion coefficient (soil-1), we can observe a very good correlation between the numerical and experimental results. In fact, rather high geosynthetic stiffness disturbs and prevents the expansion of the soil at very large deformation rates. Nevertheless, the good agreement between the numerical and analytical results demonstrates the relevance of the proposed analytical design method and the capability of the discrete element model to take account of the interactive phenomenon between the soil and the reinforcement.

2.4.2. *Numerical and experimental behavior of granular embankments reinforced with geosynthetic in areas prone to subsidence*

This work was performed in the framework of the PhD of Audrey Huckert [HUC 14] in collaboration with Laurent Briançon. Thanks to them.

The progressive development of the territory leads to the exploitation of new zones, currently neglected, because they present risks for the safety of users. This is the case of areas of potential subsidence that are related to the presence of underground cavities. The recognition techniques of these cavities generally do not make it possible to detect them exhaustively, and required filling works are often very expensive, without being able to systematically provide the guarantee of a durable solution. In recent years, the development of the geosynthetic reinforcement technique positioned at the base of the embankments has made it possible to provide economical and technical solutions for temporarily limiting the consequences of a localized sinkhole by limiting the surface deflections to acceptable values, necessary for the continuation of the traffic. The main mechanisms involved in this issue are the flexion behavior of the geosynthetic, the load transfer mechanisms within the part of the granular embankment located above the cavity and the progressive stretching of the geosynthetic obtained by friction in the anchoring zones, located on either side of the cavity.

Although significant progress had been made in recent decades to improve the design methods of reinforced geosynthetic embankments above cavities [BRI 08,

VIL 08], some scientific challenges had to be removed. The use of a discrete numerical model, coupled with full-scale experiments, has proved to be an appreciable help to clarify different assumptions such as the shape of the distribution of the vertical load acting on the geosynthetic layer or to better understand the phenomenon of expansion of the soil cylinder located above the cavity. These new results made it possible to modify the existing analytical methods towards more realistic design methods [HUC 16, VIL 16, CHA 19].

The experiments carried out in the framework of the FUI Geo-inov project [HUC 14] consist of simulating the gradual opening of a circular cavity under a granular embankment, reinforced by a geosynthetic sheet. Three experimental tests using different types of soil and geosynthetics have been realized. The instrumentation carried out made it possible to measure, at each stage of the opening process of the cavity and for each loading phase, the surface settlements, the displacements within the granular embankment, the geosynthetic strains and deflections, as well as the load transfer mechanisms within the embankment, using pressure sensors positioned at the edges of the cavity. The mechanical parameters of the granular material and of the interface between the soil and the geosynthetic were determined by laboratory characterization tests (shearing and friction tests using 0.3 m x 0.3 m boxes). We will then focus on the full-scale experiments carried out using a granular soil embankment (20/40 mm sized gravel) with a thickness h_m of 1.0 m and a PET geosynthetic consisting of a non-woven support reinforced by wires in one direction.

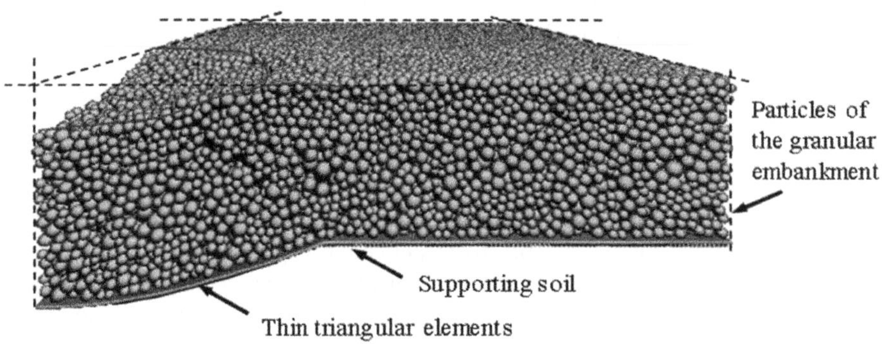

Figure 2.26. *Geometry of the reference numerical sample [HUC 14a]*

The discrete numerical model was used to take into account the large deformations in the granular embankment during expansion, shearing or compaction, to follow the evolution of the contact forces between the particles during the opening of the cavity, and to integrate the complex interaction mechanisms acting at the interface between the soil and the geosynthetic as rolling or friction. The expected results are the displacements of each particle of the granular material, the deformations and tensions of the geosynthetic layer, as well as the interaction forces at the soil–geosynthetic interface. Several geometries of the numerical model, each adapted to the diameter of the cavities tested, were used. Due to the symmetry, the cavity was positioned in one corner of the numerical sample (Figure 2.26).

The reference numerical sample used for comparison with the experimental results consists of:

– a set of discrete particles set up at a fixed porosity. Each numerical element, whose shape is close to that of the real forms of grain, consists of clumps made of two spheres of the same diameter d juxtaposed to one another. For the reference numerical model (4 m x 4 m), 32,000 elements of different sizes were used, the larger particles having, just as for the experimental material, a size twice as large as the size of the smaller particles;

– a continuous geosynthetic sheet consisting of three-node triangular elements [VIL 98]. Different tensile stiffnesses, in the direction of reinforcement and in the perpendicular direction, were considered to restore the anisotropic behavior of the geosynthetic used during the experiments. For the reference numerical sample, 12,800 triangles were used;

– a set of spheres regularly arranged and positioned at the base of the numerical model to simulate the action of an elastic subsoil. The positions of these spheres are modified to simulate the opening of the cavity. Two opening mechanisms were studied: a gradual opening of the cavity by increasing its diameter (until obtaining a diameter D, process A) and a progressive lowering of all the spheres of the supporting soil, located under the cavity for a fixed diameter cavity D (process B). For the reference numerical sample, 12,800 spheres were used to model the subsoil;

– frictionless walls positioned at the sides of the numerical sample and boundary conditions in displacement (on the horizontal directions) for the nodes sited at the periphery of the geosynthetic layer, to ensure the symmetry of the problem.

During the numerical process, the granular embankment is subjected to gravity and then to the sinkhole test by displacement of the spheres of the subsoil located above the cavity, following process A or B. The microscopic parameters of the soil particles, presented in Table 2.4, have been calibrated [SAL 09] to account for the mechanical behavior of a low-density granular material with an angle of friction at the peak of 36.5°, a friction angle at rest of 31° and a Young's modulus of 19 MPa. The interface behavior between the elements of the granular material and the geosynthetic is restored by a Coulomb-type contact law using parameters given in Table 2.4. The bidirectional behavior of the geosynthetic used in the experiments is taken into account in the numerical model using two specific tensile stiffnesses, one in the direction of reinforcement (J_x = 3,000 kN/m) and the other in the perpendicular direction (J_y = 250 kN/m). The results of the numerical model are confronted here with the experimental results considered to be the most representative. In this case and at any point of view, the numerical process is similar to the experimental protocol, namely a gradual opening of the cavity diameter up to a maximum of nominal diameter of 2.2 m.

Numerical parameters allowing us to define the granular material behavior	
Micromechanical friction parameter: $\mu = \tan(\delta)$ (-)	0.53
Normal contact stiffness: K_{nij} (N/m²)	1.0 x 10^8
Tangential contact stiffness: K_{tij} (N/m²)	1.0 x 10^8
Numerical parameters of the soil–geosynthetic interface	
Friction angle for the upper interface: $\varphi_{i\text{-}sup}$ (°)	23
Friction angle for the lower interface: $\varphi_{i\text{-}inf}$ (°)	40
Tangential contact stiffness: E_0 (MN/m³)	15

Table 2.4. *Numerical parameters used to reproduce the mechanical behavior of the granular material and the interface behavior between the soil and the geosynthetics*

In Figures 2.27 and 2.28, the surface settlements, the vertical displacements and the strain of the geosynthetic sheet, for a cavity opening diameter of 2.2 m, are compared. As can be seen in Figure 2.27, the surface settlements and the vertical displacements of the geosynthetic are fairly well reproduced by the numerical model. It can also be noted that the geometry of the numerical and experimental soil areas involved in the collapse is located mainly above the cavity. By comparison between the volumes of soil before (cylinder above the cavity) and after collapse (delimited by the surface deflection and the deformation of the geosynthetic), it is possible to estimate the coefficient of expansion of the numerical soil at 1.048, which remains quite close to the experimental value of 1.037. This result is very satisfactory if we consider that the shape of the particles and the initial density of the embankment granular material were only very roughly taken into account in the numerical modeling.

If we compare (Figure 2.28) the numerical values of the strains of the geosynthetic sheet with the values recorded by the Bragg gratings, we can also see, given the level of precision that can be achieved in situ, a fairly good match between the results. In order to analyze the load transfer mechanisms within the granular embankment, it is possible to numerically determine the vertical stresses in the vicinity of the cavity by considering all of the vertical forces acting on the geosynthetic sheet around the periphery of the cavity (annular surface defined by internal and external radii of 1.1 m and 1.5 m) and compare this value to the measurements provided by the total pressure sensors. At the end of the process of opening the cavity at D = 2.2 m, the stress increase obtained with the numerical model on the edges of the cavity is 34% for an experimental value of 49%. Again, given the potential uncertainties related to measurements (proximity to the edges of the cavity and possible tilting of the sensors), the results obtained are satisfactory.

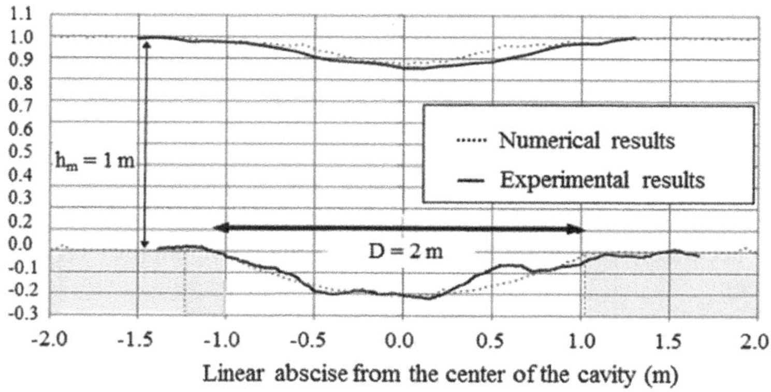

Figure 2.27. *Comparison between numerical and experimental results of geosynthetic surface settlements and deflections [VIL 16]*

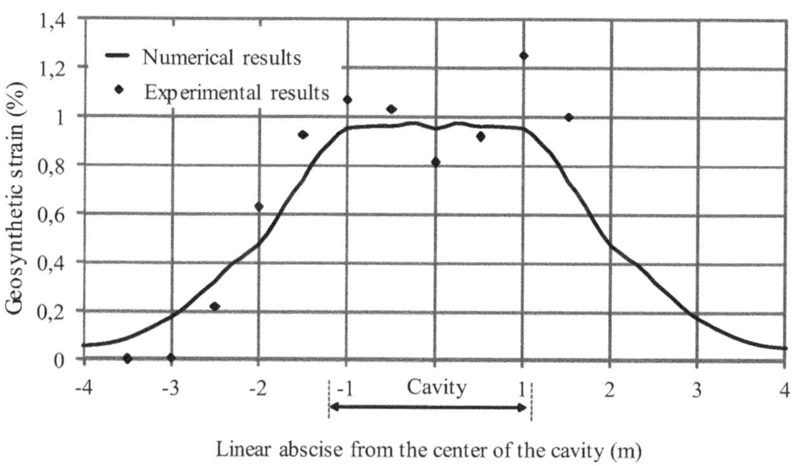

Figure 2.28. *Comparison between the numerical and experimental results of the strains in the geosynthetic sheet [VIL 16]*

The numerical tool makes it possible to measure specific quantities and to investigate complex phenomena which are difficult to quantify or to apprehend by in situ measurements. For example, it is possible to numerically determine the load transfer mechanisms that are related to a gradual change in the orientations and intensities of the interaction forces between grains of the granular embankment, following a modification of the boundary conditions. These load transfer mechanisms are highly dependent on the loading history and thus the opening mode of the cavity.

In a simple way, the efficiency of the load transfer mechanisms can be defined as the ratio between the vertical loads transmitted at the edges of the cavity and the weight w of the soil cylinder sited above the cavity. Knowing the interaction force F_g between the particles of the granular material above the cavity and the geosynthetic, it is possible to quantify the efficiency E_{ff} of the load transfer using the relation $E_{ff} = (w - F_g)/w$. Similarly, knowing the interaction forces acting at any point of the sheet, it is possible, taking account of annular areas centered on the cavity, to accurately determine the geometry of the distribution of vertical stresses on the sheet.

The influence of the opening mode of the cavity on the efficiency of the load transfers is highlighted in Figure 2.29. It should be emphasized that for this study, the size of the numerical models and the number of elements have been adapted to the cavity diameters: 0.5 m x 0.5 m x 1.0 m mesh for cavity diameters of 0.1–0.2 m (model A), mesh size of 1.0 m x 1.0 m x 1.0 m for cavity diameters of 0.2–0.5 m (model B) and meshes of 4.0 m x 4.0 m x 1.0 m for cavity diameters of 0.5–2.5 m (reference model).

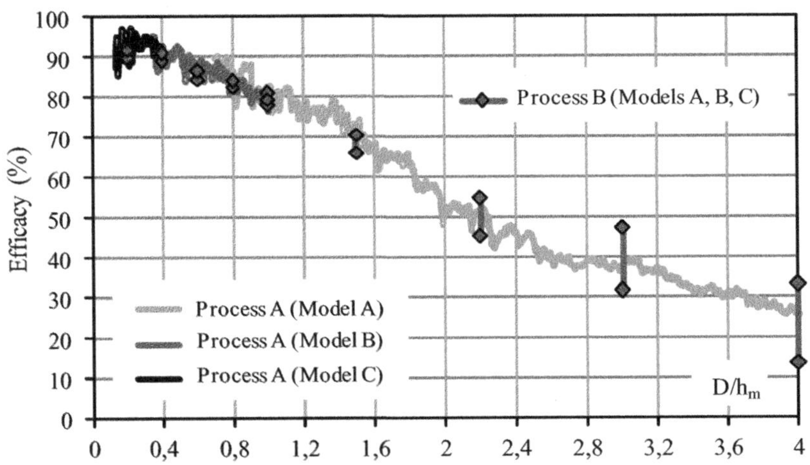

Figure 2.29. *Efficiency of the load transfer according to the ratio D/h_m for the two modes of cavity opening [VIL 16]. For a color version of this figure, see www.iste.co.uk/grange/soil.zip*

For each cavity diameter tested (process B), two points are plotted in Figure 2.29. The first characterizes the maximum efficiency of the load transfer obtained during the opening process for small deformations of the granular embankment, and the second characterizes the residual efficiency obtained after the complete opening of the cavity. The difference between these two points increases with the size of the cavity and the rate of deformation of the granular embankment. In Figure 2.29, it can also be seen that the values of the efficiency of the load transfers, from one mode of opening to the other, are quite close and whatever the cavity diameter considered (the values of the efficiency of the load transfers obtained with process A being between the two values obtained during process B).

On the other hand, the values of the surface settlements and the displacements of the geosynthetic sheet, presented for a particular value of the ratio D/h_m in Figure 2.30, are very different from one opening mode to another, which suggests that the geometry of the load distributions on the geosynthetic sheet varies depending on the mode of opening of the cavity. Moreover, the values of the expansion coefficients of the granular material sited above the cavity (1.048 for process A and 1.036 for process B) are slightly different from one opening mode to the other.

Logically, as can be seen in Figure 2.31, the expansion coefficient obtained with process A is greater, since the expanded and sheared area during the opening process is greater because of the gradual increase in the diameter of the cavity.

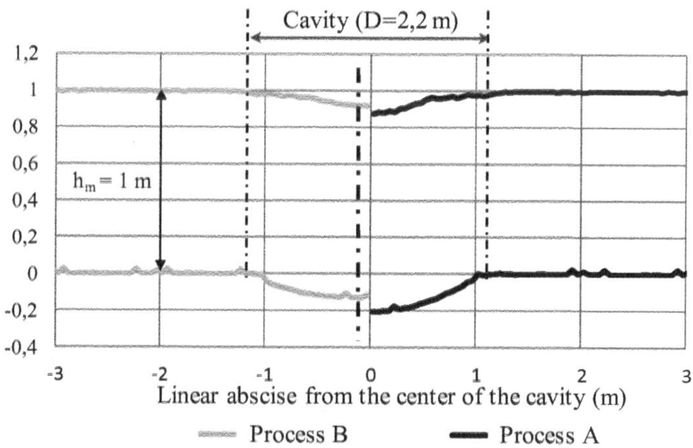

Figure 2.30. *Comparison between the surface settlements and the vertical displacements of the geosynthetic sheet depending on the opening mode of the cavity (D/h_m = 2.2 m) [VIL 16]*

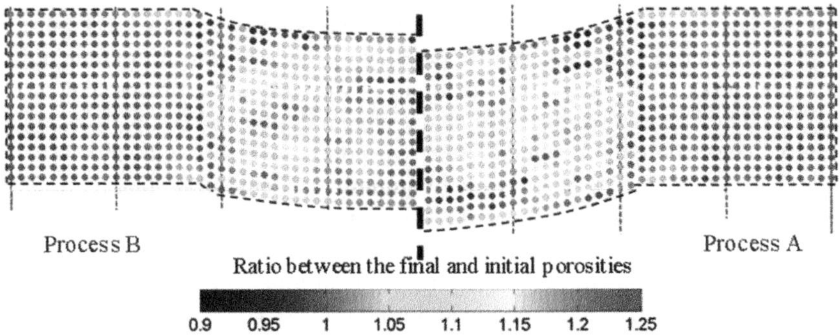

Figure 2.31. *Change in porosity within the granular embankment depending on the opening mode of the cavity (D/h_m = 2.2 m) [VIL 16]. For a color version of this figure, see www.iste.co.uk/grange/soil.zip*

To explain the difference in behavior obtained as a function of the cavity opening mode, Figure 2.32 shows the stress distributions on the geosynthetic sheet using processes A and B for $D/h_m = 2.2$. This figure shows a very clear difference in the geometry of the load distributions on the sheet, namely: a conical distribution for process A (with load concentrations in the center of the cavity) and a rather constant or inverted load distribution for process B (with a higher concentration of force in the vicinity of the edges of the cavity).

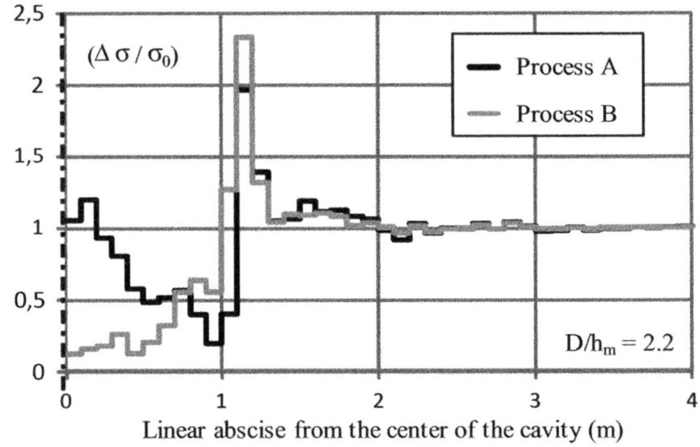

Figure 2.32. *Comparison of the geometries of the load distribution acting on the geosynthetic sheet, depending of the opening process mode [HUC 14a]*

This is due to the fact that the load transfers, towards the edges of the cavity, are systematically modified by the gradual increase of the diameter of the cavity (process A), whereas they are disturbed little throughout process B. At the end of the opening processes, the shear forces at the periphery of the collapsed soil cylinder are quite similar from one cavity opening mode to the other. This explains why the load transfer efficiencies are very close in the two cases.

The networks of the contact forces obtained, depending on the cavity opening mode for three cavity diameters, are shown in Figure 2.33. Again, it is found that the opening process B of the cavity to a fixed diameter, allows a direct transfer of loads due to the weight of the embankment to the edges of the cavity. For process A, however, these load transfer mechanisms are systematically challenged by the gradual increase in the diameter of the cavity.

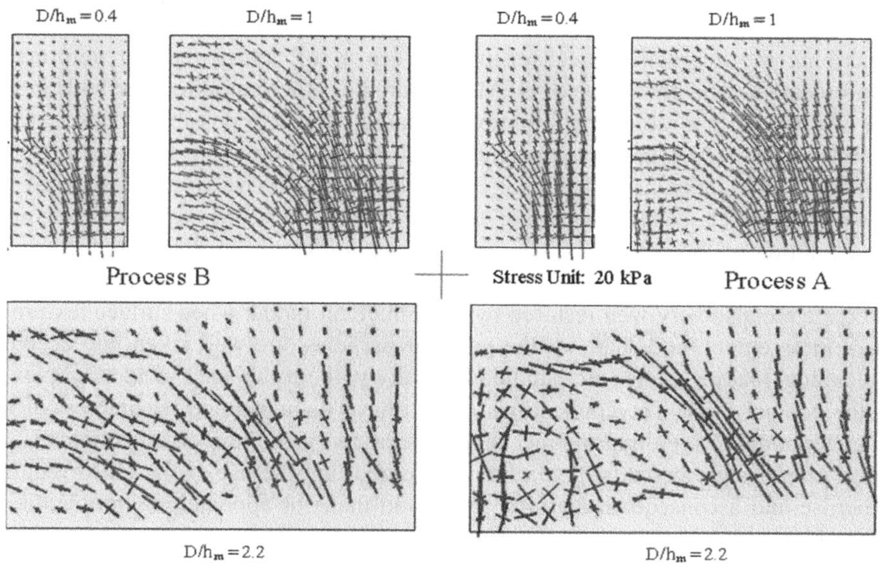

Figure 2.33. *Comparison of the contact force distributions depending on the cavity opening mode for three cavity diameters [VIL 16]. For a color version of this figure, see www.iste.co.uk/grange/soil.zip*

This study made it possible to show that, during the formation of a cavity under a granular embankment reinforced by geosynthetic, the load distribution acting on the sheet is not uniform and is largely influenced by the opening mode of the cavity. In particular, a conical load distribution, for which the load is greater in the center of the cavity, has been obtained when the diameter of the cavity increases gradually. This is related to the fact that the load transfers that take place at the beginning of the formation of the cavity, are constantly challenged during the gradual increase of its diameter. On the contrary, and as shown by the contact force networks, when the supporting soil under the embankment moves progressively, the load transfer mechanisms, towards the edges of the cavity, generated at the very beginning of the opening process, persist so that the load acting on the geosynthetic sheet is greater near the edges of the cavity. Despite these mechanisms, the overall load transferred from the collapsed zone to the stable zones seems to be not much influenced by the opening mode of the cavity. Moreover, these results also showed that the soil expansion mechanism was not homogeneous in the volume of collapsed soil and that it was strongly influenced by the opening mode of the cavity.

2.5. Conclusion

Discrete modeling has proved to be a relatively relevant numerical tool for describing the interaction mechanisms between a soil and a rigid or deformable reinforcement. The advantage of DEM compared to other numerical methods lies in the fact that, with a rather small number of parameters, it is possible to describe the complex behavior of granular materials, such as the contracting behavior of loose materials, or the expansive behavior of dense materials. The discrete nature of the numerical model makes it possible to take into account the large deformations, the evolution of the behavior of the soil following changes of densities, the swells, the collapses and other modes of rupture. Shear and expansion mechanisms at the interface are also very well restored by the numerical model when surface texture of the reinforcement, and shape of the granular particles, are well taken into account. Unloading–loading cycles are handled naturally without the need to add complexity to the contact laws. The applications treated have demonstrated the interest of the discrete model to highlight the mechanical behavior of the reinforced structures and to develop design methods. It can be emphasized that the use of the DEM requires expertise and a consequent learning phase and that, for applications in operational companies, sometimes leads to substantial computation durations.

2.6. References

[BRI 08] BRIANÇON L., VILLARD P., "Design of geosynthetic-reinforced platforms spanning localized sinkholes", *Geotextile Geomembrane*, vol. 26, no. 5, pp. 416–428, 2008.

[CHA 05] CHAREYRE B., VILLARD P., "Dynamic spar elements and DEM in two dimensions for the modeling of soil-inclusion problems", *Journal of Engineering Mechanics*, vol. 131, no. 7, pp. 689–698, 2005.

[CHA 19] CHALAK C., BRIANÇON L., VILLARD P., "Coupled numerical and experimental analyses of load transfer mechanisms in granular-reinforced platform overlying cavities", *Geotextiles and Geomembranes*, vol. 47, no. 5, pp. 587–597, 2019.

[CHE 08] CHEVALIER B., Etudes expérimentale et numérique des transferts de charge dans les matériaux granulaires – Application aux renforcements des sols par inclusions rigides, Doctoral thesis, Université Joseph Fourier, Grenoble, France, 2008.

[CHE 10] CHEVALIER B., BRIANÇON L., VILLARD P. *et al.*, "Prediction of load transfers in granular layers used in rigid inclusions technique: Experimental and discrete element method analysis", *Proceedings of the GeoFlorida 2010*, West Palm Beach, USA, pp. 1718–1726, 2010.

[CHE 11] CHEVALIER B., VILLARD P., COMBE G., "Investigation of load transfer mechanisms in geotechnical earth structures with thin fill platforms reinforced by rigid inclusions", *International Journal of Geomechanics*, vol. 11, no. 3, pp. 239–250, 2011.

[CUN 71] CUNDALL P.A., "A computer model for simulating progressive large scale movements of blocky rock systems", *Proceedings of the Symposium of the International Society of Rock Mechanics*, vol. 1, pp. 132–150, 1971.

[CUN 79] CUNDALL P.A., STRACK O.D.L., "A discrete element model for granular assemblies", *Géotechnique*, vol. 29, no. 1, pp. 47–65, 1979.

[DON 95] DONZÉ F.V., MAGNIER S.A., "Formulation of a three dimensional numerical model of brittle behavior", *Geophysical Journal International*, vol. 122, pp. 790–802, 1995.

[FUR 13] FURET A., La modélisation numérique du renforcement des sols par inclusions rigides, Engineer internship, Ecole Nationale d'ingénieurs de Saint Etienne, France, 2013.

[GIR 97] GIRAUD H., Renforcements des zones d'effondrement localisé - Modélisations physique et numérique, Doctoral thesis, University of Grenoble, France, 1997.

[GOR 13] GORNIAK J., Geosynthetic tubes filled with lightweight aggregate in new geotechnical structures: First studies, Doctoral thesis, Université Pierre et Marie Curie, Sorbonne Universités, Paris, France, 2013.

[GOR 13a] GORNIAK J., VILLARD P., DELMAS Ph., "Analytical design method for predicting the vertical loading capacity of geosynthetic tubes filled with lightweight granular material", *International Symposium on Design and Practice of Geosynthetic-Reinforced Soil Structures*, Bologna, Italy, pp. 406–415, 2013.

[GOR 15] GORNIAK J., VILLARD P., BARRAL C. *et al.*, "Experimental and analytical studies of geosynthetic tubes filled with expanded clay lightweight aggregate", *Geosynthetics International*, vol. 22, no. 3, pp. 235–248, 2015.

[GOR 16] GORNIAK J., VILLARD P., DELMAS Ph., "Coupled discrete and finite element modeling of geosynthetic tubes filled with granular material", *Geosynthetics International*, vol. 23, no. 5, pp. 362–380, 2016.

[HUC 14] HUCKERT A., Approches numérique et expérimentale du dimensionnement de renforcements géosynthétiques sur cavités et inclusion rigides, Doctoral thesis, University of Grenoble, France, 2014.

[HUC 14a] HUCKERT A., BRIANÇON L., VILLARD P., "Experimental and numerical approaches of the design of geosynthetic reinforcements overlying voids", *23rd European Young Geotechnical Engineers Conference*, EYGEC, Barcelona, Spain, pp. 133–136, 2014.

[HUC 16] HUCKERT A., BRIANÇON L., VILLARD P. *et al.*, "Load transfer mechanisms in geotextile-reinforced embankments overlying voids: Experimental and analytical approaches", *Geotextiles and Geomembranes*, vol. 44, no. 3, pp. 442–456, 2016.

[IWA 98] IWASHITA K., ODA M., "Rolling resistance at contacts in simulation of shear band development by DEM", *Journal of Engineering Mechanics*, vol. 124, no. 3, pp. 285–292, 1998.

[IWA 00] IWASHITA K., ODA M.,"Micro-deformation mechanism of shear banding process based on modified distinct element method", *Powder Technology*, vol. 109, pp. 192–205, 2000.

[JEA 92] JEAN M., MOREAU J.J., "Unilaterality and granular friction in the dynamics of rigid body collections", *Proceedings of the Contact Mechanics International Symposium*, pp. 31–48, 1992.

[JEA 99] JEAN M., "The non-smooth contact dynamics method", *Computer Methods in Applied Mechanics Engineering*, vol. 177, pp. 235–257, 1999.

[LEH 06] LE HELLO B., VILLARD P., NANCEY A. *et al.*, "Coupling finite elements and discrete elements methods, application to reinforced embankment by piles and geosynthetics", *Proceedings of Numerical Methods in Geotechnical Engineering*, Graz, Austria, pp. 843–848, 2006.

[LUD 05] LUDING S., "Shear flow modelling of cohesive and frictional fine powder", *Powder Technology*, vol. 158, pp. 45–50, 2005.

[SAL 09] SALOT C., GOTTELAND Ph., VILLARD P., "Influence of relative density on granular materials behavior: DEM simulations of triaxial tests", *Granular Matter*, vol. 11, no. 4, pp. 221–236, 2009.

[SIB 19] SIBILLE L., VILLARD P., DARVE F., ABOUL HOSN R., "Quantitative prediction of discrete element models on complex loading paths", *International Journal for Numerical and Analytical Methods in Geomechanics*, vol. 43, no. 5, pp. 858–887, 2019.

[SZA 11] SZARF K., COMBE G., VILLARD P., "Polygons vs. clumps of discs: A numerical study of the influence of grain shape on the mechanical behaviour of granular materials", *Powder Technology*, vol. 208, no. 2, pp. 279–288, 2011.

[TRA 15] TRAN Q.A., Numerical modelling of soil arching between piles, Master Mécanique, Energétique et Ingénieries, Université Joseph Fourier – Grenoble INP, France, 2015.

[TRA 19] TRAN Q.A., VILLARD P., DIAS D., "Discrete and continuum numerical modeling of soil arching between piles", *International Journal of Geomechanics*, vol. 19, no. 2, 2019.

[VIL 98] VILLARD P., GIRAUD H., "Three-dimensional modelling of the behaviour of geotextile sheets as membrane", *Textile Resarch Journal*, vol. 68, no. 11, pp. 797–806, 1998.

[VIL 08] VILLARD P., BRIANÇON L., "Design of geosynthetic reinforcements for platforms subjected to localized sinkholes", *Canadian Geotechnical Journal*, vol. 45, no. 2, pp. 196–209, 2008.

[VIL 09] VILLARD P., CHEVALIER B., LE HELLO B. *et al.*, "Coupling between finite and discrete element methods for the modelling of earth structures reinforced by geosynthetics", *Computers and Geotechnics*, vol. 36, no. 5, pp. 709–717, 2009.

[VIL 16] VILLARD P., HUCKERT A., BRIANÇON L., "Load transfer mechanisms in geotextile-reinforced embankments overlying voids: Numerical approach and design", *Geotextiles and Geomembranes*, vol. 44, no. 3, pp. 381–395, 2016.

[WAT 04] WATN A., ØISETH E., AABØE R., "Reinforced slope of road embankment with light weight aggregates", *Proceedings of the 3rd European Conference on Geosynthetics*, Munich, Germany, vol. A-5, pp. 175–179, 2004.

[WEB 66] WEBER J., "Recherches concernant les contraintes inter-granulaires dans les milieux pulvérulents", *Bulletin de Liaison des Ponts et Chaussées*, no. 20, 1966.

3

SSI Analysis in Geotechnical Engineering Problems Using a Finite Difference Method

Soil–structure interaction (SSI) problems in geotechnical engineering are often complex problems, with common interactions at different levels and between several parts of the studied problem, both under static or dynamic loading conditions. These complex conditions justify the use of adapted numerical tools to take account of the global behavior of the geotechnical structure. The presence of different materials in terms of nature and mechanical behavior calls for the need for a high modularity of the computation code used. Moreover, the geo- and structural materials usually exhibit a highly nonlinear behavior and irreversible deformations. In the family of the numerical modeling in a continuum, the finite difference method is quite well adapted to address these issues.

3.1. Introduction

The SSI problems addressed in this chapter are studied using the finite difference method, implemented in numerical models in a continuum. These methods thus belong to the family of direct methods to take account of the SSI. In this kind of method, the soil mass and the structural elements are explicitly simulated and are part of the same model. Their interactions are thus taken directly into account (contrarily to substructure or macroelement approaches). Even though the simulations can be time-consuming, this tool appears to be reliable and versatile, as well as ideal for parametric studies and for the calibration of simplified methods.

Chapter written by Daniel DIAS and Orianne JENCK.

The implementation of finite difference methods allows the resolution of stress–strain problems in a continuum medium, discretized into a finite difference mesh (numerical grid). The stress and strain tensors are determined on each point/zone of the medium. This allows the visualization of the involved phenomena (and thus their analysis), simultaneously in the ground mass and in the structural elements.

Several case studies will be addressed in this chapter, all of which are related to geotechnical engineering issues exhibiting strong SSIs: numerical studies of reinforced retaining walls, tunneling in soft ground and soft soil improvement using rigid piles, which are detailed and analyzed. However, this list of examples of SSI geotechnical problems is obviously not exhaustive. Emphasis is mainly on static loading conditions and some dynamic studies are also presented. Other types of problems are not considered in this work.

3.2. The finite difference method using an explicit scheme

Finite difference methods are numerical methods for solving differential equations with initial and/or limit conditions. Each derivative of the set of equations is replaced by an algebraic expression in terms of the variation of the field variables (stress or strain for classical geotechnical problems). These variables are defined at discrete points in the space (on the numerical grid point) and are undefined elsewhere (i.e. within the model elements). The finite difference method described here is the one implemented in the software FLAC (Fast Lagrangian Analysis of Continua) from Itasca [ITA 09], i.e. an explicit method performing a Lagrangian analysis, using a time-marching scheme and a mixed discretization procedure [CUN 82, MAR 82, BIL 93].

The time-marching scheme makes it so that even when dealing with static loading conditions, dynamic equations of motions are used in the formulation. In reality, part of the deformation energy stored by the system is converted into kinematic energy which will propagate and dissipate in the medium. In this explicit resolution scheme, the unbalance force induced in a zone will propagate in the whole model. Figure 3.1 explains the calculation sequence during one time step of duration Δt: from the forces/stresses, the Newton equation of movements is implemented to compute the new velocities and thus the displacements and strain rates. From the strain rates, the implementation of the constitutive equations gives the new stress state. On each box, each variable value is updated from the previous time step value, which remains fixed during the period Δt (i.e. explicit resolution scheme). To justify this, the time step duration should be sufficiently small to ensure that the velocity of the "calculation wave" is always higher than the physical wave propagation: two adjacent elements do not influence one another during the time lap Δt, as the information has no time to be transmitted. After a sufficient number of cycles, the information (disturbances) will propagate across the model elements, as they would propagate physically. The time step is sufficiently small if smaller than the critical time step. Instead of determining a global critical time step, which would necessitate the determination of the whole system eigenperiod (which is unpractical), the time step calculation is based on the local critical time step (with a factor of safety, as it is only an estimation), depending on the element size and element mass and stiffness (conditioning the p-wave speed). As such, stiff or small model zones may control the time step value.

This approach ensures the stability of the numerical model even if the system is physically unstable (slope sliding, for example) but necessitates that the user knows exactly what he/she is currently modeling (no black box giving the result!).

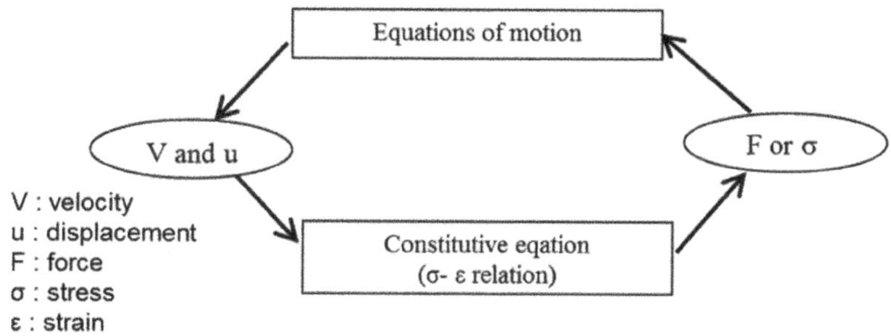

Figure 3.1. *Resolution scheme during one calculation time step*

In this type of modeling, there is no global rigidity matrix to be stored, as the equations are formulated at each time step. No iteration process is needed to compute the stresses from the strains, regardless of the constitutive model of the material. The explicit scheme is particularly well adapted for nonlinear problems, such as those encountered when dealing with SSI problems. However, the implementation of this method is counterproductive when dealing with linear problems in small deformations: in any way, the resolution will necessitate a large number of calculation cycles to reach the solution, whereas the result would be immediate if a finite element method is used.

Lagrangian analysis means that the node coordinates can be updated at each time step of the computation, the grid can thus distort with the material it represents and large deformation problems can be treated (e.g. in problems involving extensive reinforcements such as geosynthetic, in which the tension load depends on the axial strain, computed from the node positions). This is trivial because no global matrix of stiffness is used.

In this looping process, for static analysis, the movements need to be damped in order to achieve a stationary state (equilibrium or permanent flow) in a minimum number of cycles. In the described method, a damping force is applied to the model nodes (local non-viscous damping), whose direction is such that the energy is always dissipated. For dynamic simulations, more realistic damping – such as Rayleigh or hysteretic damping – should be used (see below).

In this looping process, the stationary state (equilibrium or permanent flow) is supposed to be obtained when the maximum unbalanced force on all model nodes is considered small enough, then the node velocities are very small (unless for the permanent flow case, where velocity vectors of all nodes will have a prevailing direction and will not tend to decrease during the cycling process). These aspects of convergence definition require that the user is aware of what he/she is currently doing/modeling.

In a dynamic analysis, the process of computation is the same, with the mass of the points constituting the model mesh determined from the unit mass of the surrounding zones to solve the equations of motion, relating the nodal force to the node acceleration. The damping of the waves in the soil is due to its characteristics in terms of viscosity, friction and the development of plasticity. The damping in the dynamic numerical process should reproduce the real energy loss in the system. However, this is not always possible for numerical limitations. Different sorts of damping, such as Rayleigh, hysteretic or local damping, are usually implemented. The use of local damping is simpler than the Rayleigh damping because it does not require the specification of a frequency. In order to take account of a semi-infinite medium and avoid wave reflexion in a numerical model, specific boundary conditions are assigned such as quiet (or absorbing) boundary or free-field boundary conditions. The distance to the area of interest of these latter boundaries depends on the material damping (the higher the damping, the closer the boundaries).

Moreover, it should be noted that a built-in programming language (FISH language) in the software FLAC/FLAC3D gives a high level of flexibility to the user, allowing him/her to perform the desired simulation in terms of loading conditions (load or displacement conditions, servo-control routines, etc.), to automate parameter studies, to give an unusual distribution of properties, to generate a specific grid, etc.

3.3. Application of the finite difference method to soil–structure interaction problems

It is essential to first have a conceptual idea of the problem and the estimated behavior of the global system to build a numerical model. Moreover, the questions that should arise are: could the problem be considered two-dimensional (possibly axisymmetric) or is it truly three-dimensional? What should be the size of the model (geotechnical influence zone, with a special attention when dealing with a dynamic problem) and thus what are the boundary conditions to be applied? Where are the interesting parts of the system where the mesh should be refined? What are the loading conditions (including cyclic or dynamic loading conditions)? What are the hydraulic conditions (drained or undrained conditions, hydro-mechanical coupled phenomena, i.e. the consolidation process to be taken into account)? What is the optimal constitutive model complexity, according to the level of accuracy targeted and the available material properties available, in combination with the type of loading (monotonic, quasi-static cyclic or dynamic loading)? What are the most appropriate structural elements and are interface elements between the ground mass and these elements pertinent? The response to most of these questions also determines the calculation time.

3.3.1. *Structural elements*

When dealing with an SSI geotechnical problem, structural elements are often needed. However, to represent a structural element, volumetric elements (zones, so as for the ground mass) can also be implemented with the material properties of the structure, but it can be less convenient than real structural elements such as those described below due to time calculation, which is usually higher. Moreover, the use of structural elements allows direct access to forces and moment values. For 3D analysis, typical structural elements are beam, cable, pile, shell, geogrid and liner elements (Figure 3.2). These elements comprise nodes, with a maximum of six degrees of freedom per node (three translational and three rotational components), the mass associated with the structural element being lumped at the nodes.

In the geotechnical engineering field, is it usually sufficient to consider structural elements with intrinsic linear elastic properties, sometimes with perfectly plastic yielding or plastic hinge between the elements, and in interaction with the surrounding soil with a frictional behavior (with usually a constant friction angle). Nevertheless, with the objective of achieving more advanced SSI studies, more realistic behavior of the material constituting the structure (concrete, synthetic materials, etc.) could be implemented.

Beam elements are straight, two-node finite elements – with six degrees of freedom per node (three translational and three rotational), and usually have a linear elastic behavior without failure. A collection of beam elements is used to obtain a beam structure, with the possibility of introducing a limiting plastic moment or a plastic hinge between unit elements. These types of elements are used, for example, to model framed structures or struts.

Cable elements are finite, straight elements with two nodes – with one axially oriented translational degree of freedom per node, which can yield in tension or compression, but without any bending moment resistance. A pretension load can be applied. Cables can be in frictional interaction with the grid representing the surrounding soil, to take account of grouting, for example. These types of elements are used to simulate structural elements with high tensile capacity and negligible bending effects, such as bolts and tiebacks.

Pile elements resemble beam elements, but with additional frictional interaction with the grid, representing the surrounding soil, to take account of skin-friction effects. The end-bearing effect can also be modeled. These elements are mainly used to simulate piles and rockbolt reinforcements.

Shell elements are flat elements with three nodes. Depending on the type of shell, they can either resist bending loading, membrane loading or both. A real shell structure is simulated by a collection of unit shell elements. If in contact with the surrounding soil, the elements are rigidly bound to the grid node (no relative displacement possible). These elements are implemented to simulate the structural support provided by a thin shell, when relative tangential displacements or their effects are negligible (structure slab, raft, tunnel liner, etc.).

Geogrid elements are similar to shell elements but with only membrane resistance (no flexural rigidity) and in frictional interaction with the grid representing the surrounding soil. They are used to simulate geotextiles and geogrids, which are flexible membranes in shear interaction with the soil.

Liner elements are similar to shell elements with additional frictional interaction with the surrounding ground mass (possibly on both sides of the liner). These are implemented, for example, to represent shotcrete lining or retaining walls.

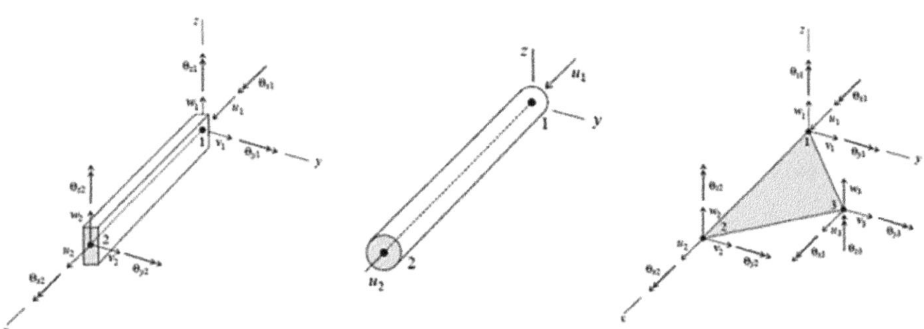

a - Beam or pile *b - Cable* *c – Shell or liner*

Figure 3.2. *Structural elements and degrees of freedom for three-dimensional modeling (source: from [ITA 09])*

In two-dimensional analysis, the classical structural elements are beam, liner, cable, pile, rockbolt and strip elements. Strip elements can yield in tension or compression but cannot bear a bending moment. The interface with the soil is defined by a nonlinear shear failure envelope. Strip elements are used in reinforced earth retaining walls. Rockbolt elements are similar to strip elements, but they can also transmit bending moments to the grid.

3.3.2. *Interfaces*

In SSI problems, interface elements are usually used between the various parts of the system (in particular between the soil mass and the structural elements, even though simulated with volume zones), in order to allow relative displacements and sliding. In three-dimensional analysis, interfaces are a collection of triangular elements (defined by three interface nodes) attached to a zone surface face, in interaction with another grid surface when in contact. In two-dimensional analysis, interfaces are segmental elements defined by two interface nodes.

In the more generalized case, interfaces have the properties of friction, cohesion, dilation, normal and shear stiffness, tensile and shear bonding (Figure 3.3). During each computation time step, the normal penetration and the relative shear velocity are calculated for each interface node and its contacting zone surface face. The interface constitutive model then gives the normal and shear forces. The interface constitutive model is usually linear (constant shear stiffness k_s and normal stiffness k_n), with a Coulomb shear strength criterion that limits the shear force according to the normal force, cohesion and friction angle of the interface. The dilation (D) causes an increase in the effective normal force when the shear strength is reached. For a bonded interface, the behavior is elastic until the tensile or shear bond strength is reached.

The effect of pore water pressure is taken into account using the effective stresses for the slip conditions.

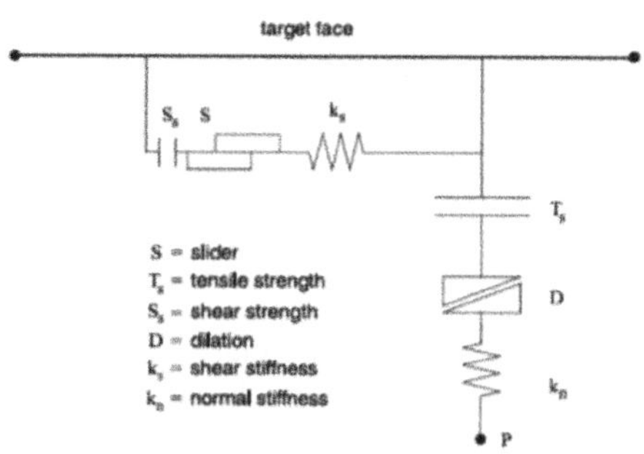

Figure 3.3. *Components of a bounded interface (source: from [ITA 09])*

For a real interface between the soil mass and a structural element, the interface properties have physical significance and should be derived from tests or data of similar materials. Concerning the stiffness values, if the physical values are very high (much more than 10 times the equivalent stiffness of adjacent zones), the computation time will be very long, as the calculation scheme performs mass scaling based on stiffness. If the normal stiffness is very small, the problem of grid interpenetration could occur. If no information is accessible on the stiffness values, or if the real values are judged as not important concerning the treated problem, the calculation duration is optimized when k_n and k_s are set to 10 times the equivalent stiffness of the stiffest neighboring zone.

3.3.3. *Constitutive models for soil*

The simplest constitutive model for soil is the linear elastic model. In the geotechnical engineering field, this should only be used when dealing with complex dynamic problems which cannot easily afford for more complex features such as plasticity, or when considering rigid structures (concrete, steel). The use of the elastic–perfectly plastic model with Mohr–Coulomb failure criteria is still very frequently used in the practice of geotechnical engineering [MES 04]. The parameters for this model have a clear meaning to all geotechnical engineers. In the field of research studies and for more advanced numerical models, more complex models are implemented, to simulate nonlinearity (sometimes even before yielding: influence of small strain stiffness), shear yielding (strain hardening or softening), compression yielding, strain accumulation, creep, etc. Some models out of the elastoplasticity framework also exist, such as hypoplastic models. To choose the most appropriate constitutive model according to the studied problem, the reader should refer to the specialized literature on the constitutive modeling of soils in the framework of the continuum mechanics [PUZ 12, HIC 08]. When using an existing code, the choice of the constitutive model is also generally restricted to the available built-in models (refer to the code user's guide), unless implementing the constitutive equations (which is feasible using the software FLAC). The choice of the complexity constitutive model should also be made in accordance with the available experimental soil data.

The influence of the complexity of the constitutive model on the global system behavior has been studied for different SSI problems, in order to determine the optimum model complexity.

[DO 13] compared the basic linear elastic–perfectly plastic model with Mohr–Coulomb failure criteria with a strain hardening constitutive model, which takes account of the hyperbolic behavior of the soil, the Cap-Yield soil (CYsoil) model, on the tunnel lining behavior and displacement field around the tunnel. They obtained a large effect of the constitutive model with higher structural forces and ground settlement when implementing the CY model.

[JAN 14] studied the face stability of shallow tunneling using a three-dimensional numerical model, implementing successively (1) the basic linear elastic–perfectly plastic model with Mohr–Coulomb failure criterion (MC model) and (2) an elastoplastic model with two isotropic hardening mechanisms (CJS2). In particular, this model allows the simulation of dilatancy before failure for dense or overconsolidated soils. Model parameter calibration was performed on triaxial and oedometric compression tests (Figure 3.4). Comparison of the tunnel face extrusion and surface settlement numerical results with experimental results clearly displayed a good agreement when using the CJS2 model, highlighting the shortcomings of the MC model for this application (Figure 3.5). The numerical modeling in the case of tunnel face bolting reinforcement was thus performed with the CJS2 model (see section 3.2.2).

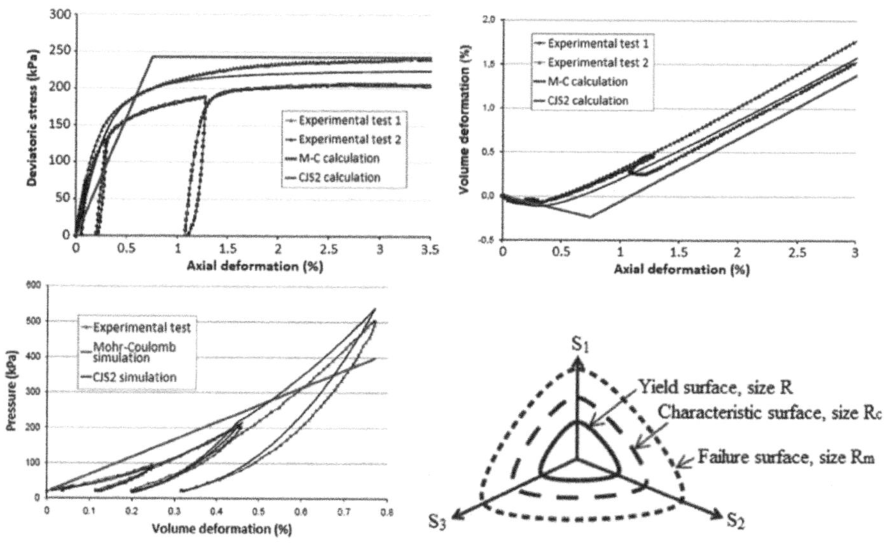

Figure 3.4. *Triaxial (top) and oedometric (bottom) test results (source: from [JAN 14]), CJS2 model surfaces in the deviatoric plane*

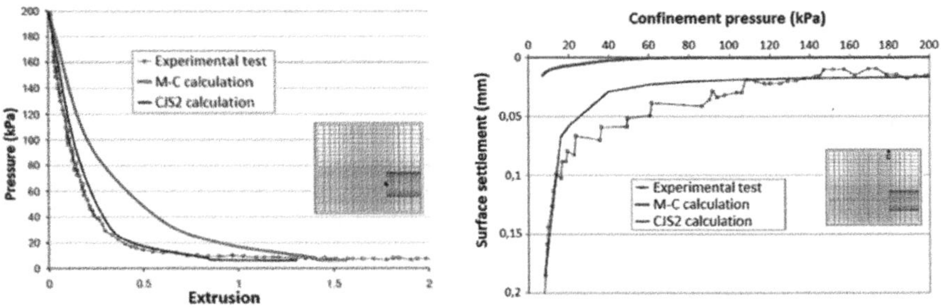

Figure 3.5. *Tunnel face extrusion (left) and surface settlement (right) due to the decrease in the pressure applied on the tunnel face (source: from [JAN 14])*

[JEN 09] highlighted that the numerical behavior of a simple piled embankment system (see section 3.4.3) under monotonic loading was almost equivalent, either implementing a constitutive model with a nonlinear elastic part and perfectly plastic failure criterion for the load transfer platform (LTP) material or using a more advanced elastoplastic model with isotropic shear hardening (CJS2 model). This is attributed to the fact that the elastic part in the first model was nonlinear: the stiffness depended on the current stress state.

[ABD 11] compared the Duncan & Chang hyperbolic model (with Mohr–Coulomb failure criterion), the elastic–perfectly plastic model with Mohr–Coulomb failure criterion and the CJS2 model for the soil modeling in the case of a mechanically reinforced retaining wall (see section 3.4.1), considering several types of reinforcing strip elements (see also section 3.4.1). They highlighted that the existence of dilatancy before failure in the CJS2 model led to higher tensile loads on the reinforcing strips at the top of the wall, which was more realistic knowing that the part of dilatancy was important when the stresses were lower. Their study also confirmed that a nonlinear constitutive model for the soil was required to appropriately take account of the shear displacement and tensile loads in the reinforcing strips, in the critical zones of the wall.

[DO 14a] studied the impact of considering soil plasticity in the case of a lined tunnel in seismic conditions (see section 3.4.2). They compared dynamic simulations performed with an elastic model to simulations completed with an elastic–perfectly plastic model. They showed that the impact of the constitutive model depends on the amplitude of the seismic signal: for a low seismic excitation, implementing a plastic constitutive model appeared useless. Nevertheless, the effect of a nonlinear elastic part and the effect of small strain stiffness were not investigated in this study.

3.3.4. *Dimension of the problem*

Simple geotechnical problems can usually be considered as two-dimensional, either with the assumption of plane strain or axisymmetric conditions. In the case of a complex SSI problem, this simplification is not always straightforward as the problem could be fully three-dimensional, in particular due to the interaction of several elements. However, despite recent advances in computing capacities, a full 3D analysis is usually complex and requires important resources, both in terms of storage and time. Some SSI studies focus on numerical processes to simplify full 3D problems in 2D calculations and highlighted the relevance or limits of 2D calculations.

Tunnel excavation is typically three-dimensional, especially around the tunnel face. [DO 17] compared 2D deconfinement methods (to take account of the displacement of the soil surrounding the tunnel before installation of the tunnel structure) to a full 3D numerical analysis (Figure 3.6), taking account of most of the components of a shield machine tunneling process. They used error functions to compare the 2D versus 3D results in terms of surface settlements, horizontal displacements on several points of the ground mass and internal forces in the lining (bending moment and normal force). They showed that the 2D "convergence confinement" method is the most appropriate to compute the soil surface settlements or structural forces in the tunnel lining, whereas the 2D "volume loss" method is more pertinent when focusing on the lateral movements of the soil.

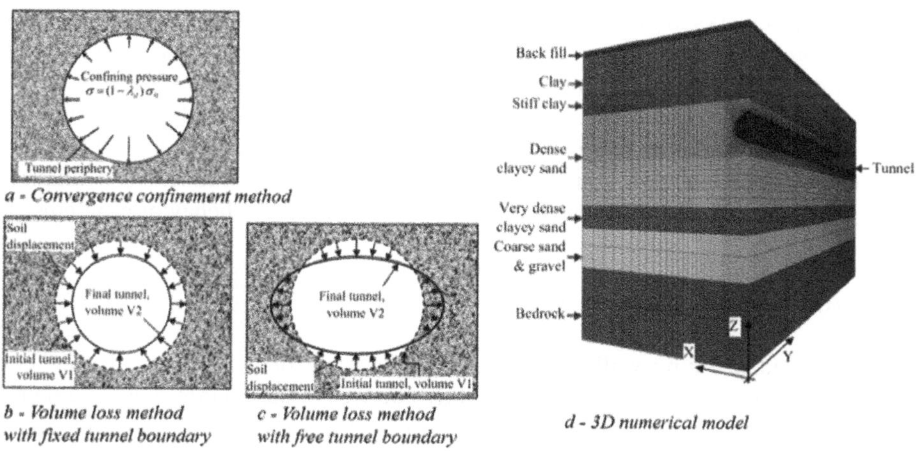

Figure 3.6. *2D and 3D numerical modeling procedures (source: from [DO 17]). For a color version of this figure, see www.iste.co.uk/grange/soil.zip*

3.3.5. *Monotonic, quasi-static cyclic and dynamic loadings*

The finite difference continuum modeling is adapted to perform static or dynamic loading simulations. Static simulations can be monotonic or cyclic (under quasi-static conditions). In the latter, an appropriate constitutive model should be used, taking into account the specific properties of the simulated materials under several unloading–reloading conditions.

The earthquake-resistant design of structures, as well as their deformations, depend on the SSIs during seismic excitation. When performing a dynamic analysis using a finite difference method, two important features should be carefully considered: the damping characteristics of the materials (in particular the soil mass) and the boundary conditions (see section 3.2), to avoid wave reflexion. Due to computation time, these types of studies are often performed with an elastic behavior of the soil, but the consideration of soil plasticity could lead to different results [LOP 17], in particular for high-amplitude seismic signals [DO 14a].

3.4. Some application examples in the geotechnical engineering field

This section relates to some selected case studies of SSI problems in the geotechnical engineering field using the finite difference method implemented in the commercial software FLAC or FLAC3D from Itasca [ITA 09]. The aim of this section is to give insight into the type of issues that can be treated, with a focus on some research works. Some additional references are given for the reader who wants to go beyond on a specific topic. Three different types of geotechnical problems are addressed in this section: reinforced retaining walls, tunneling in soft ground and rigid pile improvement. This is a non-exhaustive list of geotechnical structures involving SSI. Moreover, these examples only deal with mechanical SSI and do not treat coupled phenomenon issues such as hydro-mechanical (flow or consolidation process), thermo-mechanical (energetic geostructures) or creep behavior problems.

3.4.1. *Reinforced retaining walls*

Reinforced retaining walls comprise a high variety of techniques, all of them exhibiting strong mechanical interactions between the reinforcing elements (geotextile, geogrid, metallic or synthetic strips, etc.), the soil mass and sometimes the facing elements, leading to the effective working of the global structure. The finite difference method has been used to study the behavior of such geotechnical structures, both under static [DAM 13, YU 15a, YU 15b] or seismic loading conditions [KRI 17, CHA 17].

Mechanically stabilized earth (MSE) walls highlight strong SSI between the reinforcing strips and the soil mass, the presence of the strips improving the overall mechanical properties of the soil, due to their tensile strength and adherence or bond at the interface with the surrounding soil. [ABD 11] performed a numerical analysis, in order to better understand the behavior of such structures and contribute to the optimization of the design methods. Moreover, new types of MSE walls are reinforced with synthetic strips, generally more flexible than the traditional metallic strips, and existing design methods usually do not consider the deformation state of the structure and appear excessively conservative concerning the ultimate limit state. The stress–strain analysis, as allowed by the finite difference approach. It is therefore a powerful tool to handle this issue. The main aim of the study was to highlight the influence of the soil, the reinforcement and the soil/structure parameters on both the ultimate limit state (failure of the structure) and serviceability limit state (deformation of the structure). In order to perform an intensive parametric study, a 2D model was derived from the 3D configuration of the wall (Figure 3.7).

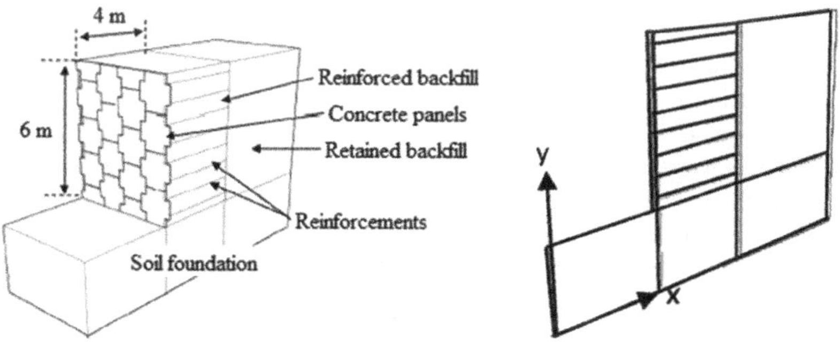

Figure 3.7. *MSE wall and 2D equivalent model (source: from [ABD 11])*

The behavior of the soil composing the reinforced backfill was successively simulated with various elastoplastic constitutive models (see section 3.3.3), the parameters being calibrated on triaxial compression test results; the reinforcement elements were simulated with strip elements (see section 3.3.1). Three types of existing synthetic or metallic strips were considered; the parameters of the interface between the strips and the soil were determined from laboratory pull-out tests, which were simulated in a numerical model dedicated to this calibration (Figure 3.8). To take account of the influence of the confining pressure on the strip/soil interface shear behavior, a model with a nonlinear shear failure envelope was used. In fact, this made it possible to consider the increase in the interface friction coefficient due to the constrained dilatancy effect, which increased the vertical stress. This feature was particularly significant at low vertical stress, i.e. at the top of the wall.

The facing concrete panels were modeled with beam elements, with interface elements to consider the frictional interaction with the backfilling soil. The construction stages of the wall were considered in this modeling.

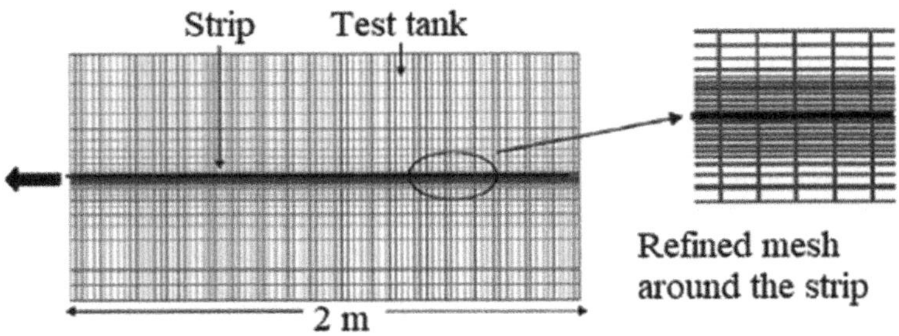

Figure 3.8. *2D numerical model of a laboratory pull-out test (source: from [ABD 11])*

The proposed numerical model first allowed a precise analysis of the failure modes and deformation of the wall (Figure 3.9). Failure was obtained by applying the c–φ reduction process to the backfill soil, which consisted of reducing the cohesion and friction angle of the soil until failure was reached (the interface friction parameter was also reduced accordingly). The factor of safety was then defined as the ratio between the initial strength and the strength at failure. The reference case was with synthetic strips (2.5 GPa elastic modulus, 100 kN/m tensile limit strength, 12% strain at tensile limit strength; initial soil/strip interface coefficient equal to 1.2 on top of the wall and 0.6 at the wall base, 0.22 MN/m²/m shear stiffness at the soil/strip interface). The factor of safety was equal to 1.51 and the maximum displacement was equal to 78 mm, which was a relatively high value as the strips presented a low stiffness. The authors mentioned that this modeling did not consider the horizontal displacement correction which could be made on a real wall construction work, achieved by varying the facing slope.

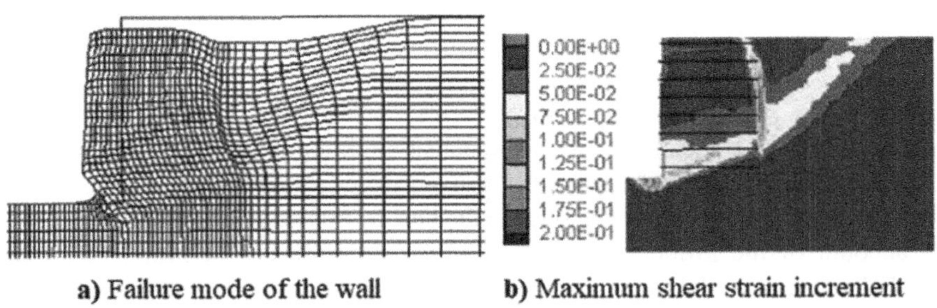

Figure 3.9. *Ultimate limit state of the reference MSE wall of 6 m height (source: from [ABD 11]). For a color version of this figure, see www.iste.co.uk/grange/soil.zip*

The parametric study performed on the soil, soil/strip interface and strip behavior parameters (Figure 3.10) showed that the soil friction angle and cohesion and the soil/strip interface friction coefficient had the most important influence on the wall factor of safety, whereas the soil cohesion and strip elastic modulus are predominant concerning the wall deformations. The soil/strip interface parameters influenced both the deformation and the limit strength of the wall, showing that a good estimation of these parameters based on experimental results seems to be essential for each type of reinforcement; this can be achieved with laboratory pull-out tests.

Parameter		Degree of influence	
		ULS	SLS
Interface	Interface shear stiffness	–	++
	Friction coefficient	+	–
Strip	Strip elastic modulus	–	+++
Soil	Young modulus	–	–
	Poisson's ratio	–	–
	Friction	+	++
	Cohesion	+	+++
	Dilatancy	–	+

Figure 3.10. *Main behavior parameters influencing the ultimate limit state (ULS) and serviceability limit state (SLS) (source: from [ABD 11])*

3.4.2. *Tunneling*

3.4.2.1. *Tunneling in soft ground and interaction with adjacent structures*

In urban areas, the construction of shallow tunnels (usually excavated using tunnel boring machines – TBMs) creates deformations and displacements of the surrounding soil which may affect existing structures and thus can be a source of unacceptable dysfunctions. A simplistic approach is to apply the greenfield deformations caused by the TBM passage to the foundations of the structure (i.e. soil movements obtained without considering the presence of the structure). However, the presence of the structure contributes to the stiffening of the system and consequently to the reduction of soil deformation and stresses in the structure. To take account of the full SSI and the tunnel excavation process, a three-dimensional model is necessary, representing the ground mass and the surface structure. The relevance of using 3D finite difference continuum modeling to take account of the tunneling process using TBM in soft soil has notably been highlighted by [DIA 13], who compared numerical results to field measurements of the Lyon subway tunnel.

An example of a numerical study of the tunneling–surface structure interaction was given by [JEN 04], where the framed surface structure was simulated by beams and shell elements. The tunnel excavation process under the structure was performed step by step and reproduced the volume loss inside the tunnel, due to the annular void between the TBM and the surrounding soil, due to its conical shape and grout consolidation (Figure 3.11). The software's built-in programming language made it possible to successfully apply the desired conditions inside the tunnel (free convergence of the tunnel walls until a limit imposed by the presence of the TBM and setting of concrete rings). The tunnel face deconfinement was not taken into consideration.

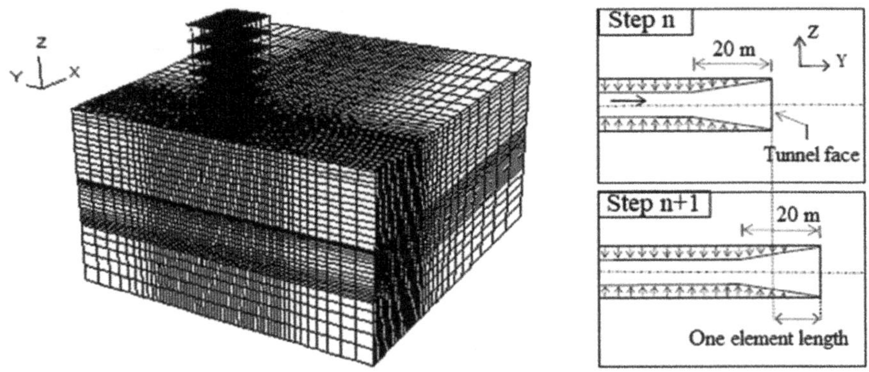

Figure 3.11. *3D numerical model (left) and sequential tunnel excavation procedure (right) (source: from [JEN 04])*

The ground mass was multi-layered, with soil properties taken from the Lisboa subway project. The soil behavior was drained and simulated with an elastic–perfectly plastic model. The surface building was simplified, and structure elements were modeled with beam elements for the columns and shell elements for the foundation raft and the slabs, so the structure had an elastic behavior (with elastic properties of reinforced concrete), which could be considered as a model limitation from a structural engineering point of view.

Several simulations varying the volume loss in the tunnel and the type of surface structure highlighted the SSI between the ground mass and the surface structure. The presence of the structure increased the surface settlement for a volume loss in the tunnel higher than 1%, in comparison to the greenfield case, and, due to the presence of a raft with a great axial stiffness and the fact that the raft is bound with the soil, the horizontal displacements were negligible under the structure in comparison to the greenfield case (Figure 3.12). Therefore, it would be conservative for the structure behavior prediction to apply the greenfield conditions to its foundations. With this full numerical model, the forces and bending moments could also be studied within the structure elements.

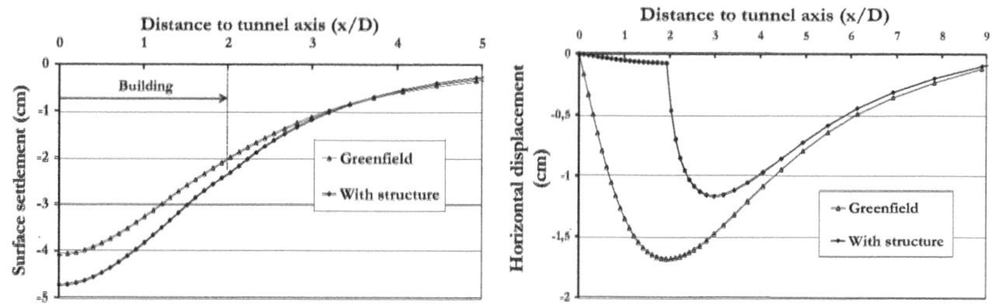

Figure 3.12. *Surface displacements (source: from [JEN 04])*

Another big issue of the interaction of tunneling with adjacent structures is the pile–tunnel interaction or piled structure–tunnel interaction. The interaction mechanisms are not fully understood yet and greatly differ considering the pile length and location, the pile group effect, the pile stiffness, etc. This problem has been addressed using finite difference numerical modeling, sometimes as a tool to calibrate simplified methods (see [LOG 01, LEE 12]).

3.4.2.2. *Tunnel face reinforcement*

In urban areas, it is necessary to control the ground movements caused by shallow tunneling, to avoid damages on neighboring structures. One of the techniques consists of preconfining the core ahead of the tunnel face with glass fiber or steel bolts. The deformation prediction of the bolted ground mass is required in order to optimize the reinforcement technique. The soil mass and the bolts are in strong interaction, and this interaction highly determines the global behavior.

[JAN 14] proposed a three-dimensional numerical modeling of the tunnel digging with bolted face, based on experimental tests performed on a centrifuge 1/50 scale down model. The modeling procedure aimed to reproduce the physical testing procedure, which was straightforward using the finite difference method as implemented in FLAC3D, as it was highly versatile. However, only half of the whole physical model placed in a rigid container needed to be simulated, due to symmetry conditions. The tunnel lining was made of steel and then assumed to be non-deformable. In the numerical model, the sliding between the lining and the soil was possible by using an interface element with a friction angle equal to two thirds of the soil mass friction angle. A latex membrane under air pressure was placed on the tunnel face, with the initial geostatic value equal to 200 kPa and the decrease in air pressure simulated the tunnel excavation, until failure occurred. In the numerical model, normal stresses were applied on the external side of the tunnel face soil zones to reproduce these conditions. The numerical model was subjected to a 50 g vertical acceleration, as in the centrifuge tests. The soil mass was Fontainebleau sand and its behavior was simulated alternatively with the linear elastic–perfectly plastic model with the Mohr–Coulomb failure criterion or with the elastoplastic CJS2 model, with the parameters being calibrated on triaxial and oedometric tests. The bolts were sand-coated PVC pipes and simulated with pile structural elements, with a frictional interface with the surrounding sand (friction angle equal to two thirds of the soil friction angle); the interface behavior properties being one of the key parameters controlling the bolting influence on displacement and tunnel face stability, as demonstrated by previous studies.

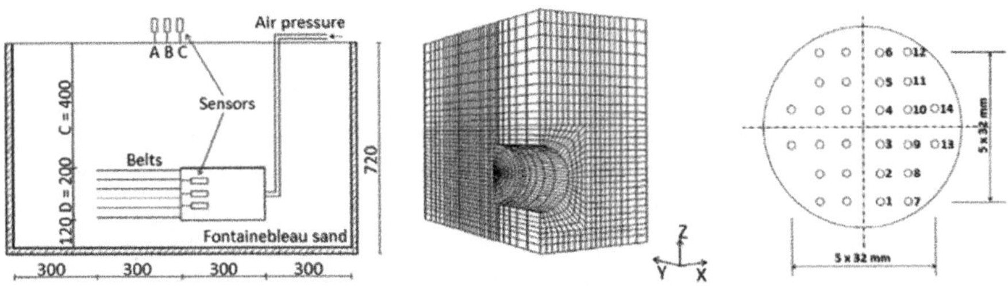

Figure 3.13. *Physical model cross-section (dimensions in mm), 3D numerical model composed of 56,000 soil zones and tunnel face bolting (source: from [JAN 14])*

The comparison between the experimental and numerical results highlighted a good agreement in terms of displacement and tunnel face stability, both in the unreinforced and bolt-reinforced cases, when using the CJS2 elastoplastic model for the soil mass behavior, as shown in Figure 3.14. Moreover, the numerical model allowed an in-depth analysis of the failure mechanisms, by representing the contour of velocity magnitude and the soil zones that had plasticized (Figure 3.15), and comparison with existing theoretical approaches could be made.

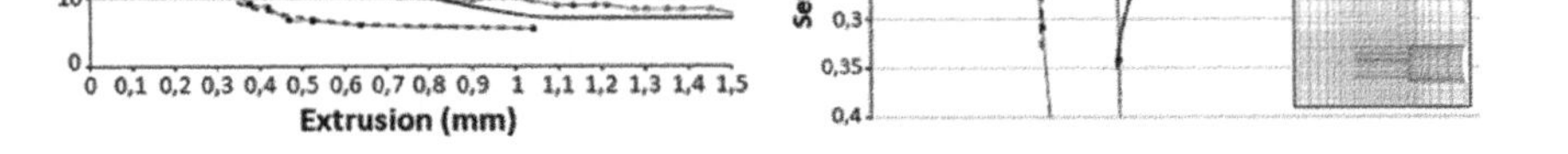

Figure 3.14. *Comparison between experimental and numerical results in terms of axial face displacement and surface settlements (source: from [JAN 15])*

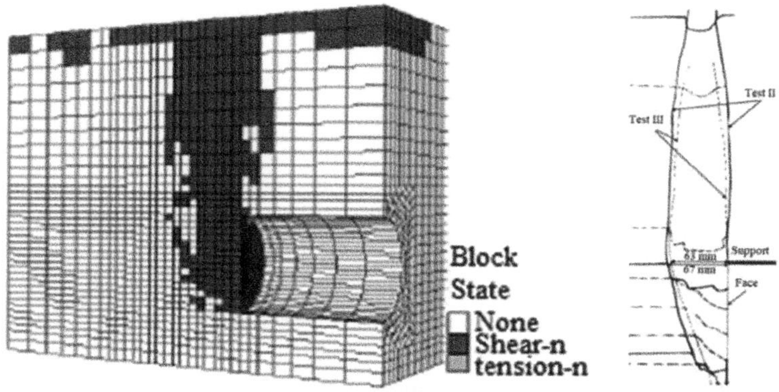

Figure 3.15. *Numerical plastic zones (left) and iso-settlement lines obtained in the centrifuge test (right) (source: from [JAN 15])*

The developed model was then used to analyze the influence of additional parameters, such as the bolting density, bolt length and bolt material. In fact, in the experimental campaign, the number of performed tests was limited.

3.4.2.3. *Tunnels under seismic loading*

The SSI phenomenon occurring in tunneling is still an open question when subjected to seismic loading. [DO 14a] studied the segmental tunnel lining behavior under seismic loading using 2D finite difference modeling (Figure 3.16), mainly in order to assess the tunnel response in terms of loads induced in the lining and to discuss the design of segmental tunnel lining. Most of the numerical studies consider seismic loads as pseudo-static load, but the literature review shows that it could lead to smaller forces in the tunnel lining than with a truly dynamic analysis. This option was thus adopted in this study, even though the calculation time was much longer.

Ground parameters (clayey sand) and tunnel features from the Bologna–Florence high speed railway line tunnel were used. Two constitutive models for the soil were successively considered: linear elastic and elastoplastic with Mohr–Coulomb failure criterion. The tunnel structure behavior was assumed to be linear elastic and simulated with liner elements, with no sliding at the ground mass interface. Segmental tunnel linings were often implemented in seismic areas, as they could undergo deformations with less or no damage compared to continuous linings, due to their flexibility. The segment elements from the tunnel lining were connected with joints, considered as elastic pins with rotational, axial and radial stiffness (Figure 3.17), whose values were calibrated on a procedure developed by [DO 14b].

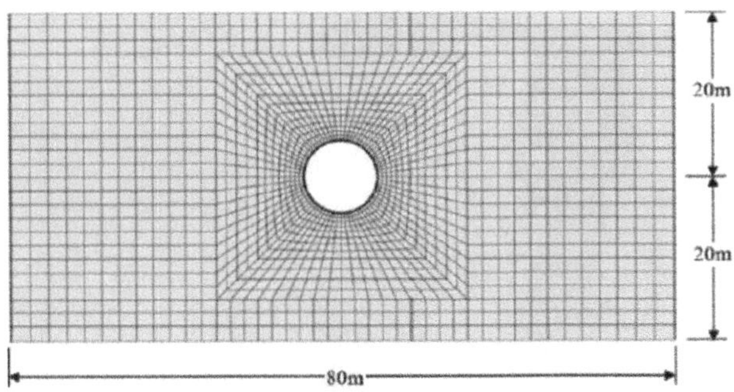

Figure 3.16. *2D plane strain numerical model (source: from [DO 14a])*

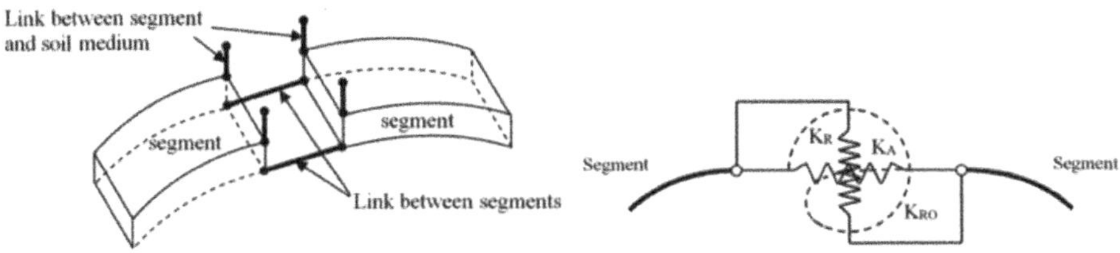

Figure 3.17. *Connection between lining segment elements (source: from [DO 14a])*

Before the dynamic loading application, the steady state of the model was obtained under static conditions (leading to initial forces in the tunnel lining). A real seismic signal with a maximum acceleration of 0.35g (or 0.0035g for comparison purposes) and a duration of 21 seconds was then given as a velocity input to the numerical model. Dynamic calculations were performed using Rayleigh damping, and quiet model boundaries were used to avoid wave reflection. Forces induced in the tunnel lining due to the seismic loading could be studied.

Even though only one single tunnel geometrical configuration and a single homogeneous ground mass was considered, a comparison with a continuous lining could be performed and bending moments reached during the earthquake were smaller in the case of segmental lining; the importance of taking into account soil plasticity under high seismic loading was underlined, as the irreversible behavior of the soil modified the tunnel loads during and after the earthquake (Figure 3.18), exhibiting the strong SSI of the studied problem; the assumption that an equivalent static solution led to smaller tunnel loads was confirmed (Figure 3.19).

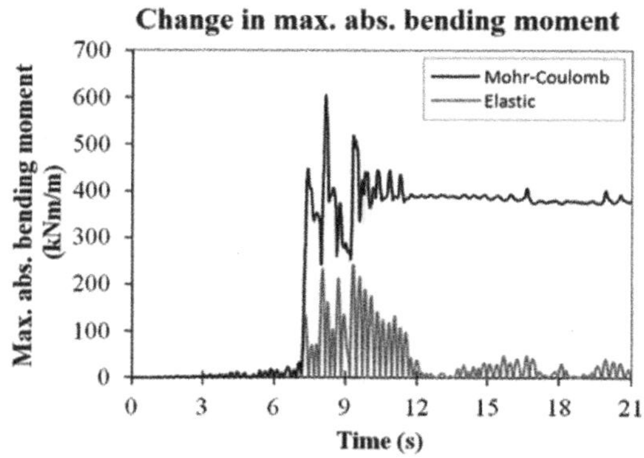

Figure 3.18. *Impact of the soil constitutive model on the bending moments in the tunnel lining (source: from [DO 14a])*

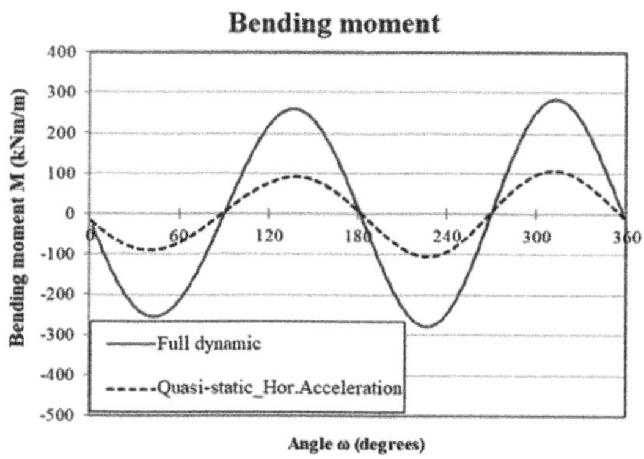

Figure 3.19. *Bending moments in the tunnel lining obtained with a dynamic calculation and a pseudo-static calculation (source: from [DO 14a])*

3.4.3. *Soft soil improvement using vertical rigid piles*

Improvement using vertical rigid piles (Figure 3.20) is one of the techniques to build structures or infrastructures on soft soil stratums. This technique is increasingly used worldwide as it is of rapid implementation and leads to large (total and differential) settlement reduction. A three-dimensional grid of rigid piles (usually unreinforced concrete) is driven through the soft soil layer down to a more competent stratum. This technique differs from those of classical piles as the piles are not structurally connected to the surface structure: a load transfer platform (LTP – usually made of granular material) is placed between the improved layer and the surface structure. An additional basal reinforcement (reinforcement geotextile or geogrid) can be placed at the platform base. Shearing mechanisms occurring both in the LTP and in the reinforced soil layer lead to the transfer of part of the surface load to the piles and to surface settlement reduction and homogenization. The different parts constituting the system are in complex mechanical interaction, making the problem difficult to handle with simple analytical methods. The numerical modeling approach using the finite difference method is one of the adapted tools to study the behavior and performance of the system, under static, cyclic and dynamic loading conditions, taking into account the behavior of each part of the system and their interactions in a unique model.

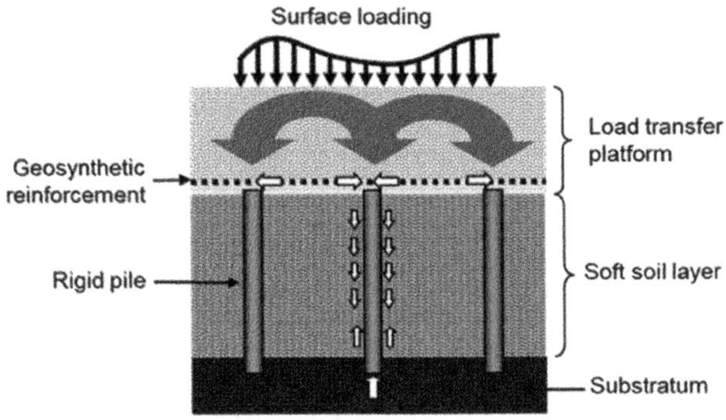

Figure 3.20. *Schematic vertical cross-section*

[JEN 07a] performed a two-dimensional preliminary numerical study, based on experimental results obtained on a laboratory small-scale model (Figure 3.21). The physical model was reduced to two dimensions and used analogical materials. Indeed, even though it is a simplification of the real system, it was developed to emphasize the mechanisms developing in the LTP, in interaction with the lower part of the system (presence of the soft soil and the piles), and to highlight the important controlling parameters by performing an intensive parametric study. Results were obtained both in terms of load distribution to the piles and the "soft soil" and displacement field in the LTP, implementing a digital image correlation (DIC) method, which is very relevant for numerical validation purposes. The main objectives of the numerical work were to assess the possibility of the modeling to reproduce the physical behavior under monotonic loading and to go further into the analysis of the load transfer and settlement reduction coupled mechanisms. In the numerical model, the piles were considered to be infinitely rigid (by fixing the nodes at the pile edges): in the small-scale model, their rigidity was actually much larger than the material simulating the soft material. However, in a non-idealized case, it is known that SSI phenomena are also governed by the relative stiffness of the various parts of the system. Interface elements were integrated to allow for relative displacements between the pile–soft material and the platform.

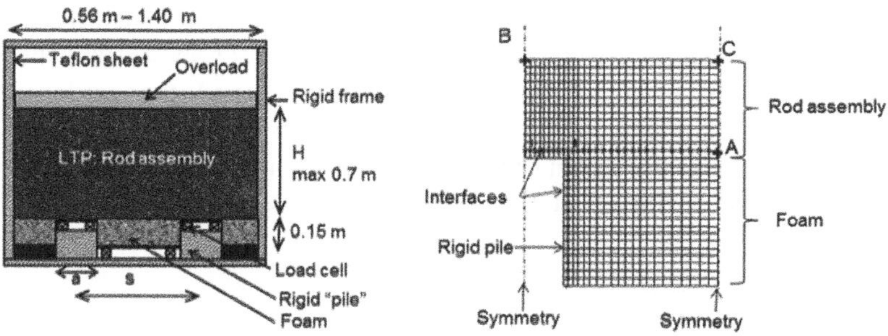

Figure 3.21. *2D schematic view of the laboratory small-scale model (left) and numerical model (one quarter of the physical model setup) (source: from [JEN 07a])*

The LTP behavior was successively simulated with an elastic–perfectly plastic Mohr–Coulomb type model using nonlinear elasticity and a two-mechanism elastoplastic model with isotropic hardening, namely the CJS2 model. The parameters were calibrated on biaxial compression tests. The foam elements representing the soft soil had a nonlinear elastic behavior, capturing exactly the results obtained with simple compression tests. The purpose was to assess the impact of the platform constitutive model complexity on the system behavior. It appeared that there was no big difference between results obtained with both models, mainly because nonlinear elasticity was considered for the elastic–perfectly plastic model. The comparison with the experimental results showed a good agreement in terms of displacements in the whole model; however, the numerical model overestimated the proportion of the load which was transmitted to the pile through the LTP by developing the arching effect. This divergence has been overcome by implementing a numerical model using the discrete element method (DEM) [JEN 09a]. In fact, in this approach of modeling, the soil particles are simulated by discrete elements, so the soil mass can largely deform at the pile head corners, due to soil element rearrangement, whereas this aspect is hidden in a continuum approach. An intensive numerical parametric study was undertaken to highlight the role of the geotechnical parameters, both in the continuum and discrete modeling. Both approaches showed that the shear strength of the LTP material (macroscopic friction angle and apparent cohesion for the continuum model, friction coefficient and bond between particles in the DEM) was the preponderant feature governing the load distribution and settlement reduction.

A major interest in performing numerical modeling based on experimental results is its possibility to obtain additional information, not accessible in the experiment, as the stress distribution in the LTP. This can help the development of simplified methods by analyzing more precisely the morphology of soil arching within the LTP, identified as the main phenomenon of load redistribution and settlement homogenization on the LTP surface. Figure 3.22 shows that the mean principal stresses are oriented towards the pile head and the earth pressure coefficient, i.e. the ratio between horizontal and vertical stresses is equal to 0.4 on top of the pile and 2.3 at the crown of the vault (corresponding to the reaching of the limit equilibrium states with a friction angle equal to 24°).

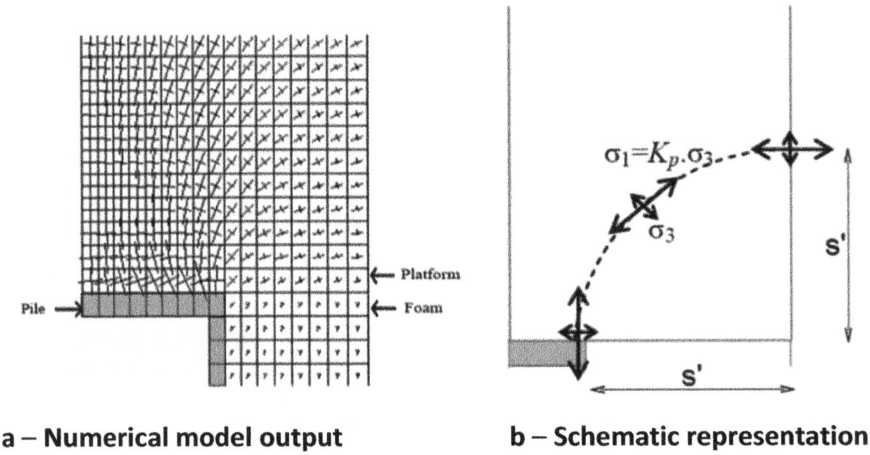

Figure 3.22. *Principal stress orientation around the pile in the continuum model (source: from [JEN 09a])*

Further work was undertaken by performing full three-dimensional analysis on a realistic case, using the same numerical modeling strategy [JEN 09b] and the modified Cam-clay model for the soft soil. Although not compared and validated on experimental data, this work proposes a numerical procedure to model piled embankments. In this modeling, an infinitely rigid substratum and rigidly embedded piles are considered, by fixing the node of the lower model boundary. A rigid pile grid usually contains tens or hundreds of regularly placed piles; then, due to symmetry conditions, only one quarter of a pile and its zone of influence has been simulated (Figure 3.23): the grid points of the vertical boundaries were fixed in the out-of-plane horizontal direction. One of the objectives was to perform parametric studies on various parameters. In particular, the impact of the soft subsoil compressibility has been highlighted, in interaction with the embankment properties (Figure 3.24), whereas the subsoil compressibility is currently not taken into account in simplified design methods. As for the 2D analysis, the dominant impact of the embankment material shear strength on the load transfer has been highlighted [JEN 07b]. Another aim was to assess the behavior of a current embankment section including lateral slopes (Figure 3.25), where the piles are then subjected to horizontal loading and bending moments, which are important aspects to be taken into account in the pile design.

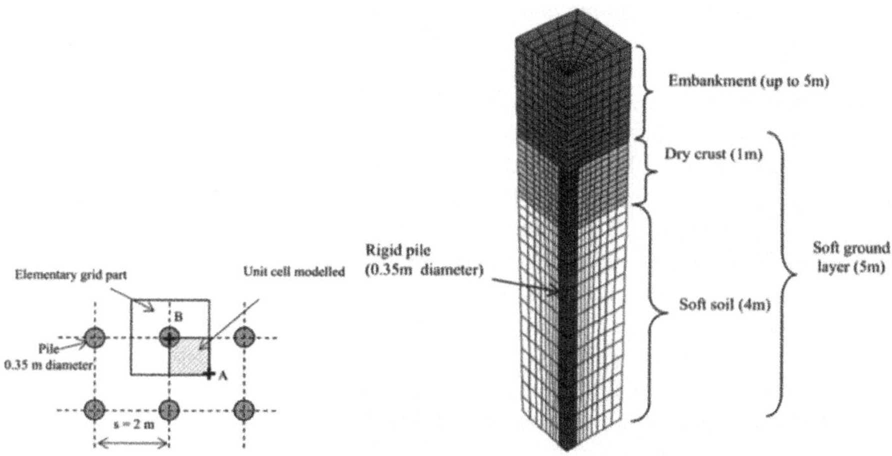

Figure 3.23. *3D numerical model of the pile grid unit element (source: from [JEN 09a])*

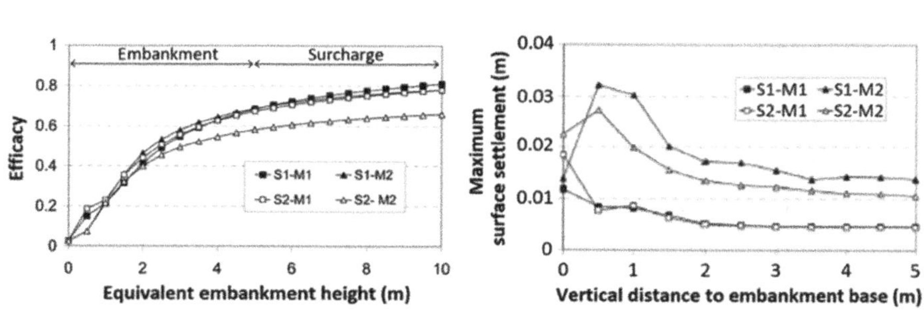

Figure 3.24. *Parametric study results for two soft deposit compressibilities (S1 and S2, with S1 being more compressible than S2) and two embankment materials (M1 and M2, with M1 being stiffer and having a higher shear strength than M2) (source: from [JEN 09a]). The depicted surface settlement is the maximum value due to the application of the next 0.5 m-thick layer*

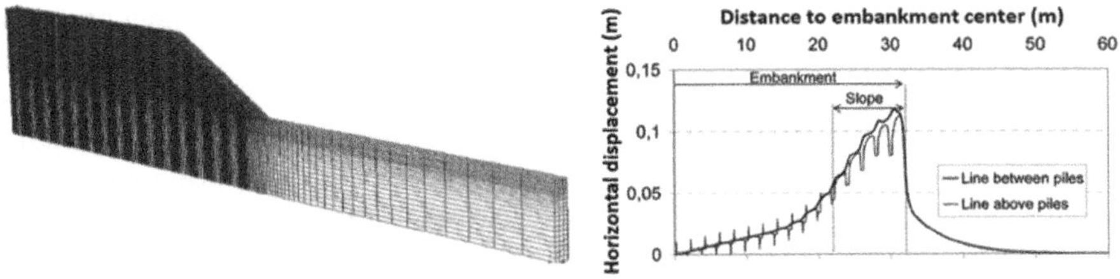

Figure 3.25. *3D numerical model of an embankment current section and horizontal displacements at the top of the piles (source: from [JEN 09a])*

The question of the relevance of simulating the load transfer onto the piles by the 3D arching effect in the embankment material using a continuum approach as the finite difference method has been addressed by [TRA 19], by comparing numerical modeling results in a continuum (FDM) and implementing the DEM (Figure 3.26). In the continuum approach, the embankment material behavior was simulated with the cap yield model to take account of friction hardening and softening. The comparison of both approaches is addressed in detail in Chapter 5 of this book. Some discrepancy between discrete and continuum models was observed when there was a relative sliding between the granular layer and the pile cap in the DEM. In fact, the compatibility conditions in the continuum model prevented this sliding and induced a stress concentration at the pile corner. Moreover, the continuum approach was not able to take account of porosity change that occurred in the DEM for a loose embankment material. This was due to the simplicity of the constitutive model used.

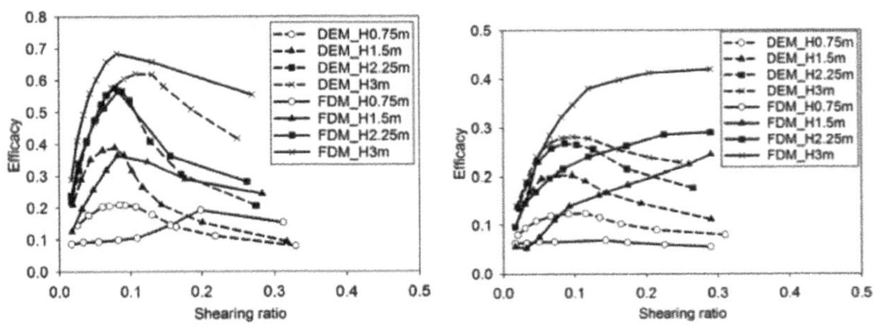

Figure 3.26. *System efficacy according to the embankment material shear ratio for a dense (left) and a loose (right) material, for several embankment heights (source: from [TRA 19])*

Even though a realistic case of a piled embankment is considered, one major limitation of this modeling is the lack of validation on experimental results. Thus, [NUN 13] performed a numerical study of full-size experiments at the Chelles

(France) test site (Figure 3.27). The test site was a 5 m-high embankment built on 8 m-thick compressive alluvial soil. The site was divided into four zones: three with pile reinforcement (two also with horizontal geosynthetic reinforcements – geotextile or two layers of geogrid) and one unimproved, to obtain the reference behavior.

Figure 3.27. *Chelles experimental site (source: from [NUN 13]). For a color version of this figure, see www.iste.co.uk/grange/soil.zip*

The performance of the embankment support system was assessed by monitoring data (total stresses, horizontal and vertical displacements). This fully instrumented study was ideal for validating the finite difference numerical modeling approach, also as the site geotechnical characterization allowed the model parameter calibration. The lateral friction limit of the piles was determined using an instrumented axially loaded test performed on the site on a floating pile, which allowed the input of the interface element parameters between the piles and the soft ground layer in the numerical model. The site instrumentation indicated a rapid consolidation of the alluvial soft soil, and most of the deformations were recorded during the embankment construction, justifying a numerical modeling performed in drained conditions.

Two numerical models were considered: an elementary cell, considering only one pile, and a global model, simulating the whole experimental site and interaction between adjacent zones (Figure 3.28). To fully take the global system behavior into account, several parts of the system were explicitly simulated: the several layers of soft soil, the piles, the interface between the piles and soil, the platform and/or the embankment, the geosynthetic layer(s) when present and also the substratum (sandy gravel layer) underneath the reinforced soft layer into which the piles are embedded. The global model comprised 445,000 volume elements (using a coarser grid than for the elementary cell). The platform and/or the embankment behavior were simulated with an elastic–perfectly plastic model with Mohr–Coulomb failure criterion, as well as the substratum, the soft layer behavior was represented by the modified Cam-clay model (see the model parameters in Table 3.1) and the piles had a linear elastic behavior. The calculations were performed in large strain mode, in order to activate the membrane effect in the geosynthetic layers, simulated with geogrid elements.

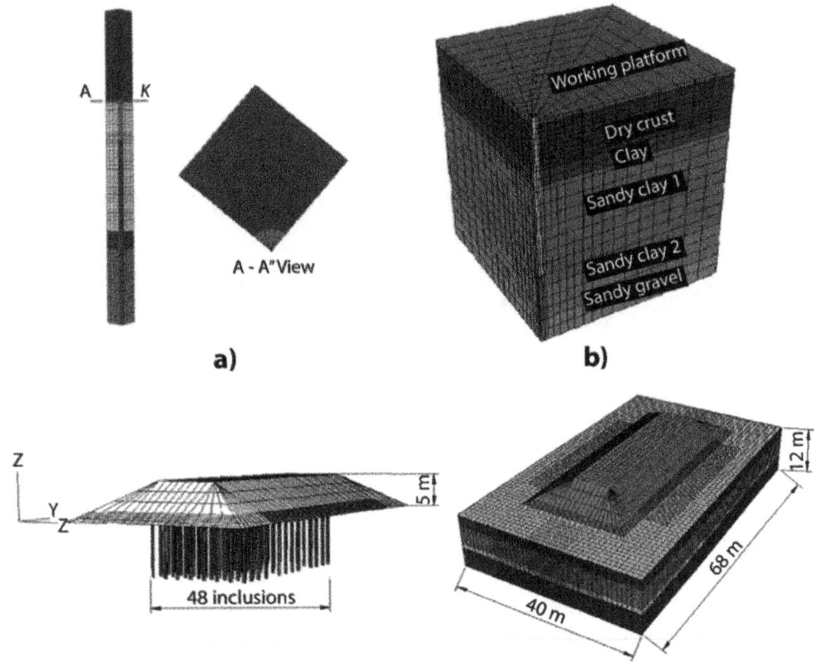

Figure 3.28. *3D numerical models of the Chelles experimental site: elementary cell, model for pile loading test and global model (source: from [NUN 13]). For a color version of this figure, see www.iste.co.uk/grange/soil.zip*

Parameter		Embankment fill	Load transfer platform	Silty clay (dry crust)	Clay	Sandy clays 1	Sandy clays 2	Sand and gravel alluvial deposits
Density, γ	[kN/m³]	19.1	21	20	15	20	20	20
Thickness, Δz	[m]			1.7	0.6	4.2	1.5–2.5	
CC/(1 + e0)	[-]			0.1	0.2	0.06	0.08	
Compressibility indices,								
CC	[-]			0.2	0.54	0.1	0.13	
λ	[-]			0.087	0.235	0.044	0.056	
Swelling indices,								
CS	[-]			0.03	0.05	0.01	0.01	
κ	[-]			0.013	0.022	0.005	0.004	
Void ratio, e0	[-]			1	1.7	0.7	0.6	
Friction ratio, φ'	[°]	36.6	36	26	26	26	26	33
Cohesion, c'	[kPa]	17.3	61	0	0	0	0	0
Dilatancy, ψ'	[°]	6.6	3				3	
Overconsolidation pressure, $\sigma'p$	[kPa]			30				
Pressuremeter modulus, EM	[MPa]							34.5
Pressuremeter limit pressure, Pl	[MPa]							2.3
Elastic modulus, E	[MPa]	50	70					76.6
Poisson ratio, ν	[-]	0.3	0.3					0.3

Table 3.1. *Model parameters (source: from [NUN 13])*

The first numerical validation step was the comparison of the results obtained on the non-reinforced zone and on the embedded pile loading test, to validate the modeling hypothesis of the soft soil layers and an embedded pile behavior (including shaft and toe resistance). Comparisons were then performed for the reinforced zones, in terms of stress efficacy (proportion of the total load transmitted to the piles; Figure 3.29) and settlements (Figure 3.30). The proposed numerical model (in particular, the global model, taking into account the influence of adjacent zones) made it possible to obtain results close to the site instrumentation results, showing the ability of the numerical modeling to reproduce the phenomena leading to the system performance. A comparison with simplified analytical methods showed that the system performance was underestimated, as they did not take into account all the features of such a complex system (in particular the soil compressibility and the pile–soil skin friction). However, some discrepancies between experimental and numerical results were underlined for some cases. In particular, the role of the geosynthetic reinforcement could not be accurately taken into account.

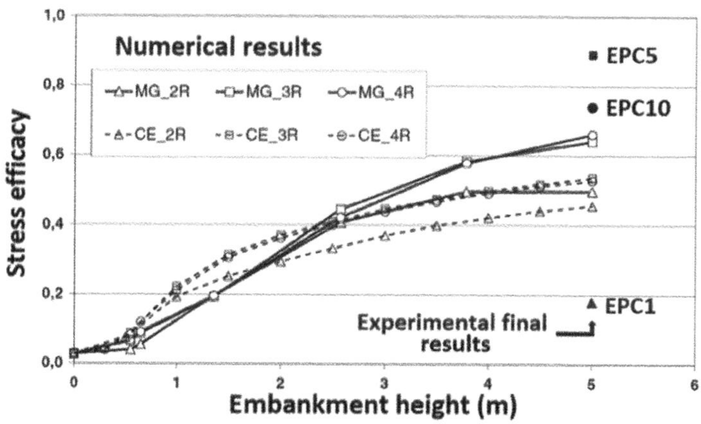

Figure 3.29. *Comparison of the numerical (CE = elementary cell; MG = global model) and experimental results in terms of stress efficacy (source: from [NUN 13])*

Another instrumented case study of soil improvement with rigid piles and comparison to a finite difference numerical modeling was reported by [BRI 15]. The project consisted of constructing an industrial building on a thick compressible clay layer: the hard sandstone bedrock is located 50 m down the soil surface. In this case, soil improvement consisted of 1 m-in-diameter-piles installed along a 6 m square grid and a 2 m-thick granular platform located between the piles and the surface 1.5 m-thick concrete slab to ensure the load transfer. As the aptitude of the thin alluvial layer located between the clay layer roof and the head of the piles was not verified to ensure a good compaction of the LTP material, 0.5 m of the alluvial layer was removed and replaced by a cement gravel layer. A geotextile was placed around the pile at 0.5 m to limit the hanging effect of the treated soil layer on the piles. The work site was largely monitored: a total of 100 sensors were installed in the LTP, soft soil and concrete piles.

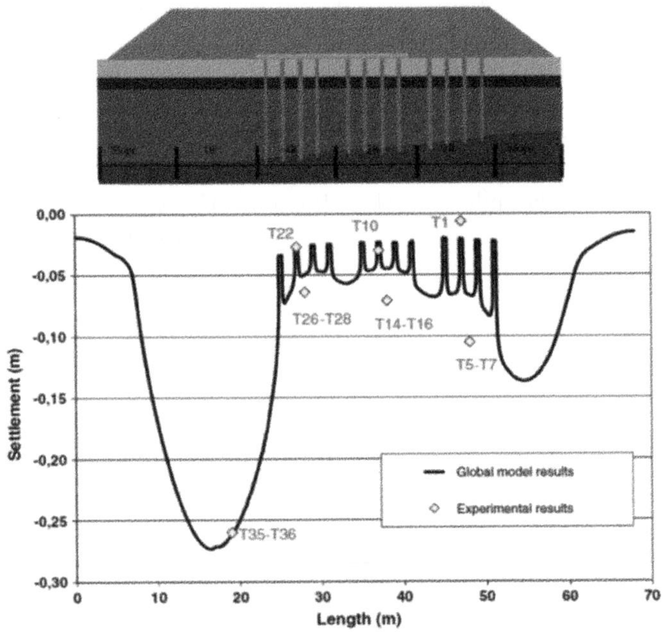

Figure 3.30. *Comparison of the numerical and experimental results in terms of settlements (source: from [NUN 13]). For a color version of this figure, see www.iste.co.uk/grange/soil.zip*

The finite difference numerical model comprised all the elements described above (Figure 3.31), from the hard bedrock up to the surface concrete slab, on which an additional loading was applied up to 150 kPa (vertical stress applied by the building and exploitation loadings). On the site, the grid contained 292 rigid piles, covering an area of 3,624 m². Only one unit cell was represented in the numerical model, considering a current grid rigid pile. The soil and the pile were modeled by volume elements, and interface elements were considered along the pile, with a Mohr–Coulomb failure criterion. For the pile–soil interfaces, an adherence (cohesion) value and a null friction angle were used, to obtain a constant friction limit. The effect of pile installation was not taken into account: the vertical piles and the soft soil were set up in only one phase, which constituted the initial state, and static equilibrium under self-weight was reached.

The numerical procedure then reproduced the successive work stages, with calculation until static equilibrium at each stage. The calculation was performed in drained conditions and corresponded to the observed conditions of quasi-instantaneous pore pressure dissipation. The clay deposit was composed of five successive clay layers, and their behavior was simulated by the modified Cam-clay model. Due to the lack of data on its constitutive behavior, an elastic–perfectly plastic model with a Mohr–Coulomb type failure criterion was used to simulate the granular material, the treated soil, the alluvia and the bedrock behavior. The slab and the piles were made of concrete and were assumed to behave elastically. The numerical model consisted of approximately 5,436 zones and 6,508 grid points. The time requested for the FLAC3D analysis of the model was about 15 hours when using a 2.40 GHz core i7 CPU computer.

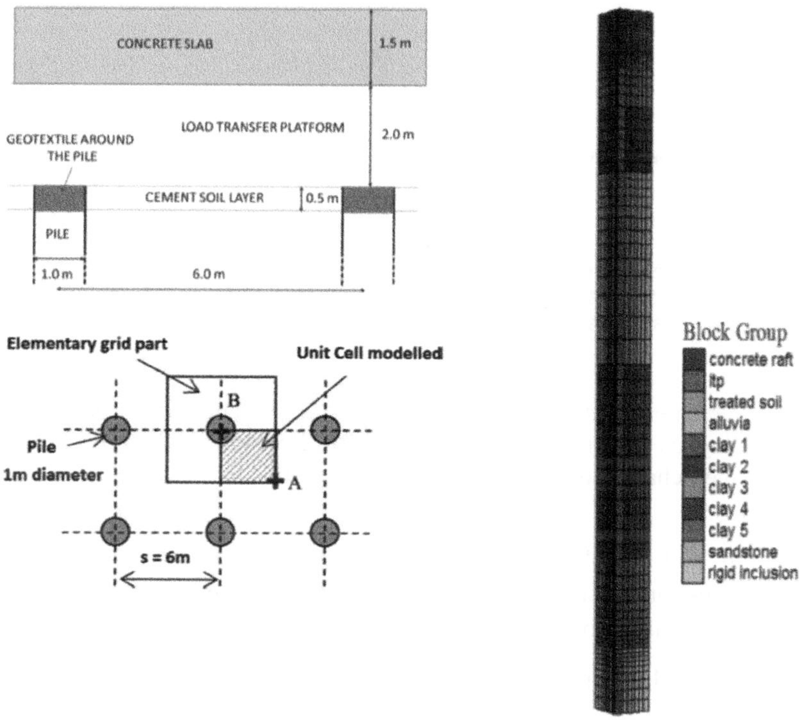

Figure 3.31. *Schematic cross-section of the upper part of the improved system (left, top), top view of the pile grid (left, bottom) and numerical model (right) (source: from [BRI 15]). For a color version of this figure, see www.iste.co.uk/grange/soil.zip*

Numerical results were compared to experimental measurements in terms of stress distribution at two levels in the platform: at the base and at mid-height (to assess the load transfer within the platform), differential settlement in the platform and pile strain, at various stages of the construction process. For all construction stages, the numerical results concurred with the experimental measurements (Figure 3.32). The numerical model appeared to be able to effectively represent the mechanisms developed inside the LTP, in interaction with the pile–soft soil system, so the behavior of the system has been predicted for the additional 150 kPa storage load. The numerical models then provided a stress value equal to 1 MPa on the pile head, an average stress equal to 110 kPa on the soft soil surface and a differential settlement equal to 0.3 mm just under the concrete slab. These results are very useful for validating or optimizing the pile and concrete slab design.

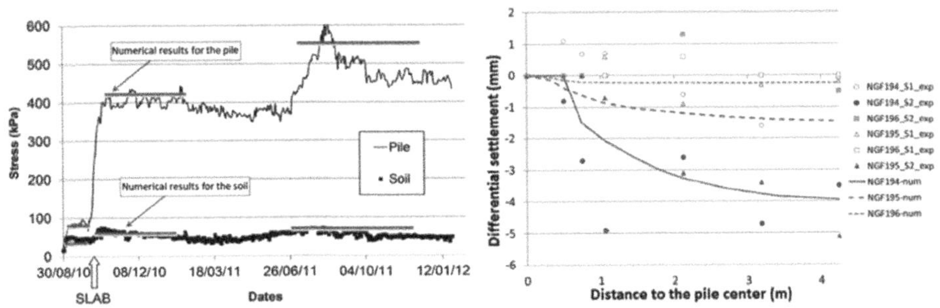

Figure 3.32. *Stress at the platform base (left) and differential settlement at various heights in the platform (source: from [BRI 15]). For a color version of this figure, see www.iste.co.uk/grange/soil.zip*

The good agreement between *in situ* measurements and finite difference numerical modeling, as shown by [NUN 13] and [BRI 15], justifies the use of this latter to predict the behavior of future pile-improved works, but it is influenced by several features:

– good quality and complete geotechnical site investigation, in order to give relevant input values for the stratigraphy and for the constitutive model parameters, and to assess the hydraulic conditions (drained or undrained conditions);

– pile load test realization on the same site, in order to assess the pile–soil interface behavior;

– implementation of a numerical model taking explicitly the various parts of the system into account, including interfaces at the pile shaft and the presence of the substratum and, if necessary, global modeling of the site;

– use of appropriate constitutive models: an elastic–perfectly plastic model with a Mohr–Coulomb failure criterion for the platform and the modified Cam-clay model for the compressible layer seem appropriate for monotonic loading. The structural parts (piles, surface slab) can behave simply elastic (influence of their stiffness);

– implementation of a numerical modeling procedure reproducing the successive working stages. However, it seems that not considering the pile installation process is not a limitation.

The finite difference model presented previously could quite truly represent a pile improvement system under monotonic loading. However, in some applications, quasi-static cyclic loading is applied on the system, such as for storage tanks and heavy load storage areas. Under such loading conditions, the durability of the mechanisms leading to the performance of the system is still an open question and no guideline exists yet. To address this issue, [HOU 17] performed experiments on a laboratory small-scale model and the numerical modeling back analysis of the behavior under cyclic loading was then completed using the finite difference modeling approach. The laboratory model consisted of a rigid box of 1 m^2 square section, containing 20 rigid piles (Figure 3.33). The scale reduction on the length compared to a real site was 1/10. Around the piles, the box was filled with a compressible material with $C_c/(1+e_0)$ equal to 0.34, on a thickness equal to 0.4 m, to represent the soft soil layer. A granular platform (LTP) composed of granular material is placed on top of the pile-compressible layer system, on a thickness equal to 50 or 100 mm. A uniform vertical load (surface loading) was applied on the surface, either directly using a soft membrane under pressure covering the entire model surface or by first placing metallic plates at the platform surface, to examine the case of a rigid slab on the surface. Low frequency vertical cyclic loading was studied by applying pressure-controlled cycles. The system was instrumented with load and displacement sensors, and four semi-piles were placed along a model border, two of them behind a Plexiglas window, allowing for photographs to be taken during the experiments and then obtaining the displacement field using a digital image correlation (DIC) method. This allowed for the analysis of the mechanisms developing in the granular platform and their evolution during the cyclic loading, and for the comparison with the numerical modeling approach, in addition to the results given by the instrumentation, located in the central part of the setup. This experimental study had some shortcomings compared to a real pile-improved structure, mainly due to the fact that the similitude between small scale and real scale was not entirely satisfied (the model has been scaled on the length, but not on the stress level, as it is under normal gravity), and because analogue material was used. So, the results cannot be directly extrapolated to a real system. Nevertheless, it constituted a powerful tool for validating numerical modeling approaches and their ability to represent the behavior observed physically, on a laboratory model where the experimental conditions were well controlled.

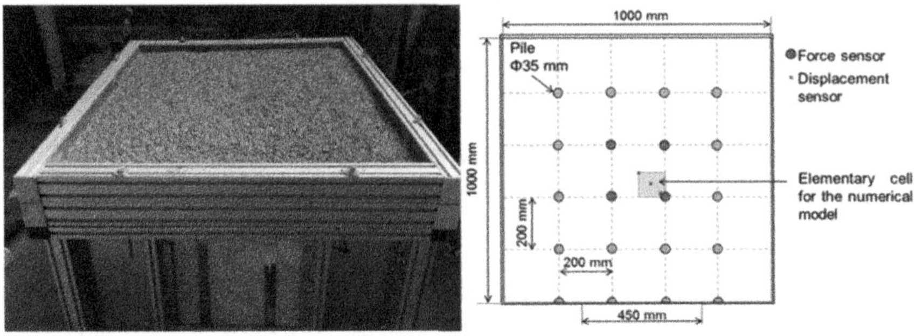

Figure 3.33. *Photograph of the laboratory small-scale model and schematic horizontal cross-section at the pile head level. For a color version of this figure, see www.iste.co.uk/grange/soil.zip*

The finite difference numerical model (Figure 3.34) resembles the model previously presented in [JEN 09b], with the dimensions of the 3D small-scale model. As the surface loading was uniform, only one quarter on a unit pile grid cell was represented. If explicit paths of loading–unloading are considered, it is usually recommended using an elastoplastic model with kinematic hardening (which is able to take account of plastic deformation under unloading of the cyclic loading history). For repeated cyclic loading (traffic load type), an equivalent viscoplastic constitutive model can be considered, but the precise behavior under the loading–unloading cycles cannot then be studied. In this first numerical study, an elastoplastic with isotropic hardening for the shear mechanism with a nonlinear elastic part (cap yield model, available as a built-in model in FLAC3D) was implemented, in order to assess the pertinence of this modeling to reproduce the observed behavior, before considering a more complex constitutive model. The parameter was determined based on cyclic triaxial tests performed on the LTP granular material. The soft material behavior was simulated with the modified Cam-clay model and parameters were calibrated based on an oedometric test result. The pile was represented by volume elements with elastic properties of the pile material. Interface elements were placed between the various parts of the system. In order to reduce the computation time, which can be high due to the high number (50) of loading–unloading cycles, a numerical procedure implicitly taking the presence of the rigid slab was developed. In fact, the presence of rigid structural elements, in contrast to the soil stiffness, considerably increases the computation time. The presence of the rigid slab on the surface was taken into account by imposing a uniform vertical displacement on the grid points on top of the LTP and a routine controlled the corresponding average vertical stress on the whole surface, to reproduce the experimental loading procedure.

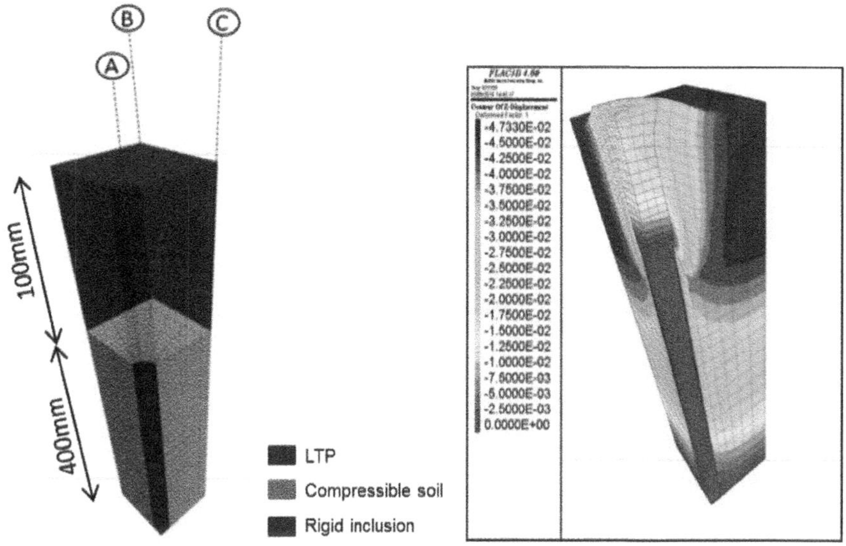

Figure 3.34. *Numerical model (left) and vertical displacement field in the model for a monotonic loading (30 kPa vertical stress applied on the surface), without rigid slab. For a color version of this figure, see www.iste.co.uk/grange/soil.zip*

The numerical and experimental results were first compared under monotonic loading and a good agreement was obtained both in terms of load transmitted to the piles and displacement field in the LTP (comparison could be made using the results obtained by the DIC method applied in the laboratory tests), for both surface boundary conditions (uniform vertical stress or uniform settlement). The results obtained under cyclic loading are shown in Figure 3.35 (the efficiency is the proportion of the total vertical load which is transmitted to the pile heads). The numerical back analysis of the tests performed in the laboratory showed that the numerical model was able to effectively reproduce some observed features, such as the basal settlement accumulation measured during the cyclic loading and the load efficiency of the system for the case without rigid slab. These encouraging results could be obtained even with an elastoplastic with isotropic hardening mechanism model, due to the nonlinearity of the model combined with the stress redistribution in the LTP during the cycles. Nevertheless, the proposed model was not able to accurately simulate the load reparation in the LTP in the case of the rigid slab on the surface: the numerical modeling underestimated the efficiency; this could be attributed to the non-explicit numerical modeling of the rigid slab and/or to the punching of the piles through the LTP that might have occurred in the experimental tests. Moreover, the settlements recorded under the post-cyclic loading were overestimated by the numerical model, as the constitutive model used was not able to take account of the soil rigidifying process under cyclic loading.

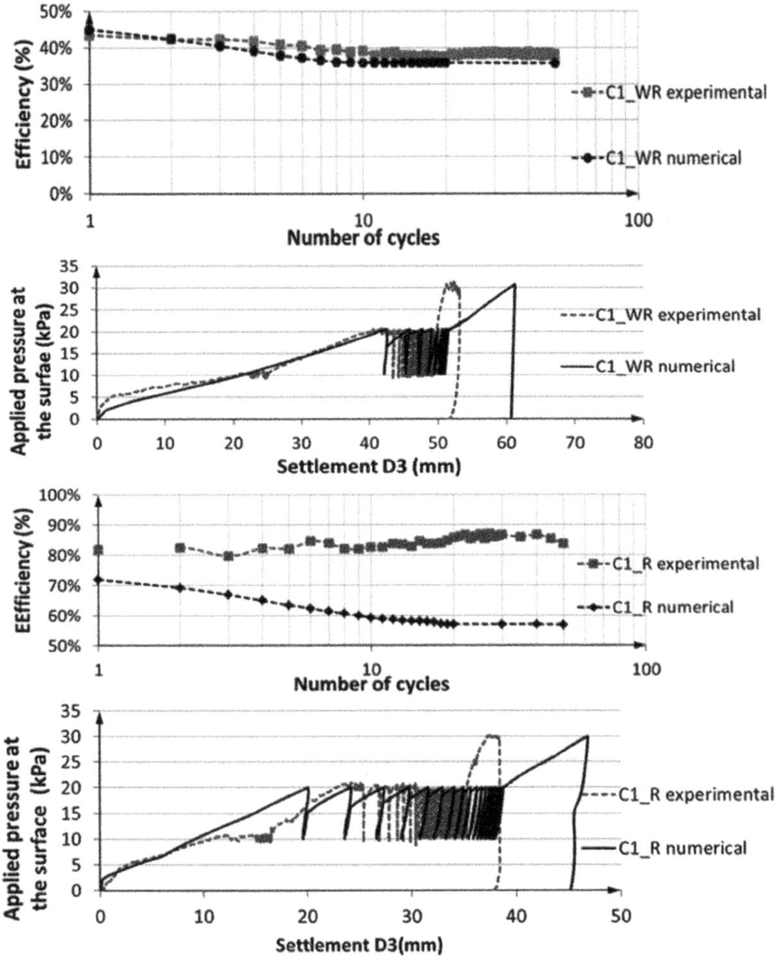

Figure 3.35. *Comparison of experimental and numerical results in terms of load efficiency and settlement at the LTP base, in the center of the model, for the cases without rigid slab (WR) and with rigid slab (R) (reference state in after model construction, just before applying the external load) (source: from [HOU 17]). For a color version of this figure, see www.iste.co.uk/grange/soil.zip*

In addition, the proposed numerical modeling can give access to additional information, useful in the practice, and difficult to obtain in the experimental part, as the surface total and differential settlements and the surface stress distribution under the slab, required a design of the structures that we will place on the improved system. Moreover, additional configurations can be simulated. For example, the influence of the surface slab stiffness: only the two extreme cases were investigated experimentally (infinitely soft or infinitely stiff), but the study demonstrated the strong influence of the surface boundary condition, so the strong interaction between the improvement system and the type of structure placed on the surface. This study would nevertheless necessitate taking explicitly this part of the system into account by implanting structural elements (shell elements) or volume elements with the corresponding slab stiffness – even if this would considerably increase the computation time. Moreover, the implementation of a more complex constitutive model for the LTP would be required to take the observed behavior into account even better.

Concerning the behavior of the piled structure under seismic loading, only a few studies have considered the nonlinearity of the soil response and most have limited to a single pile. [LOP 17] performed a dynamic numerical analysis using FLAC3D (Figure 3.36). Linear elastic and elastoplastic constitutive models were used to represent the soil behavior and different support (pile head connection and embedment) conditions were taken into account. The dynamic loading was a sinusoidal acceleration wave applied at the base of the model in the horizontal direction. Lateral absorbent boundaries were placed to minimize wave reflection. Rayleigh damping for the soil and local damping for the superstructure and rigid elements were used. The analysis of the internal forces and the displacement in the vertical reinforcements during the dynamic loading was performed for the several cases studied. The embedment conditions of the piles affected the shear forces and bending moments (in particular at the toe level), which were higher in piles anchored in the bedrock compared to the case of piles placed on the bedrock or floating piles, due to extra kinematic forces induced by the lateral deformation of the surrounding soil. Compared to the case of the piles connected to the upper structure, the system with rigid inclusions allowed the reduction of the head level efforts, due to the presence of an LTP between the structure and the rigid inclusions. In fact, for the connected pile system, the connection with the structure increased the inertial forces at this level.

Another study of the seismic ground response in the presence of rigid inclusions performed using the finite difference method was presented in [MÁN 16]. It used a viscoelastic model for the soil and showed that the dynamic characteristics of the structure at the surface were the main parameters affecting the ground response of the system.

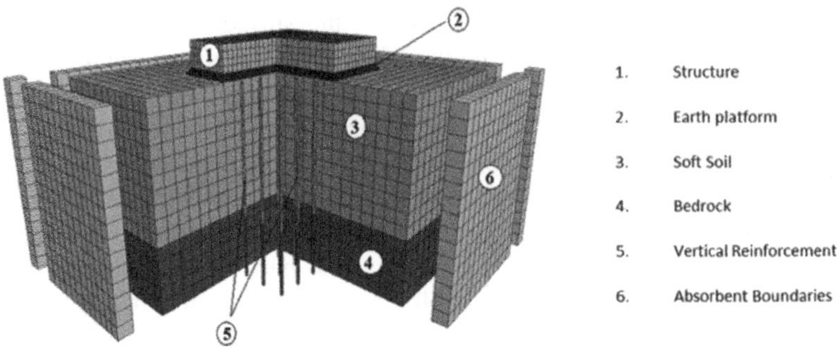

Figure 3.36. *Numerical model (source: from [LOP 17]). For a color version of this figure, see www.iste.co.uk/grange/soil.zip*

3.5. Conclusion

The finite difference numerical method addressed in this chapter is one of the direct methods to study the SSI problem in the field of geotechnical engineering. It allows the explicit modeling of both the ground mass and the structural element in the same model in a continuum, taking the interaction between the various parts fully into account. However, it necessitates the accurate modeling of the behavior of several parts and their interfaces; thus, it also necessitates the knowledge of the appropriate material properties.

In this chapter, it was shown that this type of numerical modeling is an interesting and efficient tool to help for the development or improvement of design methods, for current or new types of geotechnical structures. For this purpose, the comparison with experimental results is compulsory (using physical modeling and/or field testing). In particular, by easily achieving parametric studies, the numerical modeling method makes it possible to clearly identify the controlling parameters, which should be inevitably taken into account even in simplified design methods.

However, in the field of geotechnical engineering, this method is adapted to accurately predict the behavior of geotechnical structures, both in terms of ultimate and serviceability states, by giving access to the stress and displacement fields throughout the structure. Nevertheless, this necessitates the preliminary accurate knowledge of the stress–strain behavior of the constituting materials, given by an appropriate site investigation and interpretation. The implementation of numerical models also makes it possible to optimize the envisaged structure, by simulating various possibilities.

3.6. References

[ABD 11] ABDELOUHAB A., DIAS D., FREITAG N., "Numerical analysis of the behaviour of mechanically stabilized earth walls reinforced with different types of strips", *Geotextiles and Geomembranes*, vol. 29, pp. 116–129, doi:10.1016/j.geotexmem.2010.10.011, 2011.

[BIL 93] BILLAUX D., CUNDALL P., "Simulation des géomatériaux par la méthode des éléments Lagrangiens", *Revue française de géotechnique*, vol. 63, pp. 9–21, 1993.

[BRI 15] BRIANÇON L., DIAS D., SIMON C., "Monitoring and numerical investigation of a rigid inclusions – Reinforced industrial building", *Canadian Geotechnical Journal*, vol. 52, pp. 1–13. dx.doi.org/10.1139/cgj-2014-0262, 2015.

[CHA 17] CHAUDHARY B., HAZARIKA H., KRISHAN A.M., "Effect of backfill reinforcement on retaining wall under dynamic loading", in HAZARIKA, H., KAZAMA, M., LEE, W.F. (eds), *Geotechnical Hazards from Large Earthquakes and Heavy Rainfalls*, Springer Japan, 2017.

[CUN 82] CUNDALL P.A., "Adaptive density-scaling for time-explicit calculations", *4th International Conference on Numerical Methods in Geomechanics*, Edmonton, pp. 23–26, 1982.

[DAM 13] DAMIANS I.P., BATHURST R.J., JOSA A. *et al.*, "Comparison of finite element and finite difference modelling results with measured performance of a reinforced soil wall", *Geo Montréal*, 2013.

[DIA 13] DIAS D., KASTNER R., "Movements caused by the excavation of tunnels using face pressurized shields: Analysis of monitoring and numerical modeling results", *Engineering Geology*, vol. 152, pp. 17–25, available at: http://dx.doi.org/10.1016/j.enggeo.2012.10.002, 2013.

[DO 13] DO N.A., DIAS D., ORESTE P. *et al.*, "3D modelling for mechanized tunnelling in soft ground – Influence of the constitutive model", *American Journal of Applied Sciences*, vol. 10, no. 8, pp. 863–875, doi: 10.3844/ajassp.2013.863.875, 2013.

[DO 14a] DO N.A., DIAS D., ORESTE P., "2D seismic numerical analysis of segmental tunnel lining behaviour", *Bulletin of the New Zealand Society for Earthquake Engineering*, vol. 47, no. 3, pp. 206–216, 2014.

[DO 14b] DO N.A., DIAS D., ORESTE P. *et al.*, "2D tunnel numerical investigation: The influence of the simplified excavation method on tunnel behaviour", *Geotechnical and Geological Engineering*, vol. 32, no. 1, pp. 43–58, doi: 10.1007/s10706-013-9690-y, 2014.

[DO 17] DO N.A., DIAS D., "A comparison of 2D and 3D numerical simulations of tunnelling in soft soils", *Environmental Earth Sciences*, vol. 76, no. 3, 2017.

[HIC 08] HICHER P.Y., SHAO J.F., *Constitutive Modeling of Soils and Rocks*, ISTE Ltd, London and John Wiley & Sons, New York, 2008.

[HOU 17] HOUDA M., JENCK O., EMERIAULT F., "Rétro-analyse numérique du comportement d'un massif de sol renforcé par inclusions rigides sous chargement cyclique (Numerical back analysis of the behaviour of soft soil improved by rigid piles under cyclic loading)", *Proceedings of the International Conference on Soil Mechanics and Geotechnical Engineering*, pp. 12–18, Seoul, September 2017.

[ITA 09] ITASCA CONSULTING GROUP., Flac 3D Version 4.0, User's guide, 2009.

[JAN 14] JANIN J.P., DIAS D., "Tunnel face reinforcement by bolting – Numerical modelling of centrifuge tests", *Soils and Rocks*, vol. 37, no. 1, 2014.

[JEN 04] JENCK O., DIAS D., "Analyse tridimensionnelle en différences finies de l'interaction entre une structure en béton et le creusement d'un tunnel à faible profondeur", *Géotechnique*, vol. 8, no. 54, pp. 519–528, 2004.

[JEN 07a] JENCK O., DIAS D., KASTNER R., "Two-dimensional physical and numerical modeling of a pile-supported earth platform over soft soil", *Journal of Geotechnical and Geoenvironmental Engineering*, vol. 133, no. 3, pp. 295–305, doi: 10.1061/(ASCE)1090-0241(2007)133:3(295), 2007.

[JEN 07b] JENCK O., DIAS D., KASTNER R. *et al.*, "Three-dimensional finite-difference modeling of a piled embankment on soft ground", *Proceedings of the 5th International Workshop on Application of Computational Mechanics in Geotechnical Engineering*, Guimares, Portugal, 2007.

[JEN 09a] JENCK O., DIAS D., KASTNER R., "Discrete element modelling of a granular platform supported by piles in soft soil – Validation on a small scale model test and comparison to a numerical analysis in a continuum", *Computers and Geotechnics*, vol. 36, no. 6, pp. 917–927, doi: 10.1016/j.compgeo.2009.02.001, 2009.

[JEN 09b] JENCK O., DIAS D., KASTNER R., "Three-dimensional numerical modeling of a piled embankment", *International Journal of Geomechanics*, vol. 9, no. 3, doi: 10.1061/(ASCE)1532-3641(2009)9:3(102), 2009.

[KRI 17] KRISHNA A.M., BHATTACHARJEE A., "Behavior of rigid-faced reinforced soil-retaining walls subjected to different earthquake ground motions", *International Journal of Geomechanics*, vol. 17, no. 1, 2017.

[LEE 12] LEE C.J., "Three-dimensional numerical analyses of the response of a single pile and pile groups to tunnelling in weak weathered rock", *Tunnelling and Underground Space Technology*, vol. 32, pp. 132–142, 2012.

[LOG 01] LOGANATHAN N., POULOS H.G., XU., K.J., "Ground and pile-group responses due to tunnelling", *Soils and Foundations*, vol. 41, no. 1, pp. 57–67, 2001.

[LOP 17] LOPEZ JIMENEZ G.A., DIAS D., OKYAY U.S. *et al.*, "Dynamic behavior of the soil-pile-structure interaction in soft soils. Influence of the pile type", *International Journal of Geomechanics*, April 2017.

[MÁN 16] MÁNICA-MALCOM M.A., OVANDO-SHELLEY E., BOTERO JARAMILLO E., "Numerical study of the seismic behavior of rigid inclusions in soft Mexico City clay", *Journal of Earthquake Engineering*, vol. 20, no. 3, pp. 447–475, 2016.

[MAR 82] MARTI J., CUNDALL P.A., "Mixed discretization procedure for accurate solution of plasticity problems", *International Journal of Numerical and Analytical Methods in Geomechanics*, vol. 6, pp. 129–139, 1982.

[MES 04] MESTAT, P., BOURGEOIS, E., RIOU, Y., "Numerical modelling of embankments and underground works", *Computer and Geotechnics*, vol. 3, pp. 227–236, 2004.

[NUN 13] NUNEZ M.A, BRIANÇON L., DIAS D., "Analyses of a pile-supported embankment over soft clay: Full-scale experiment, analytical and numerical approaches", *Engineering Geology*, vol. 153, pp. 53–67, available at: http://dx.doi.org/10.1016/j.enggeo.2012.11.006, 2013.

[PUZ 12] PUZRIN A., *Constitutive Modelling in Geomechanics*, Springer, Berlin, 2012.

[TRA 19] TRAN Q.A., VILLARD P., DIAS D., "Discrete and continuum modelling of soil arching between piles", *International Journal of Geomechanics*, vol. 19, no. 2, available at: https://doi.org/10.1061/(ASCE)GM.1943-5622.0001341, 2019.

[YU 15a] YU Y., BATHURST R.J., MIYATA Y., "Numerical analysis of a mechanically stabilized earth wall reinforced with steel strips", *Soils and Foundations*, vol. 55, no. 3, pp. 536–547, 2015.

[YU 15b] YU Y., DAMIANS I.P., BATHURST R.J., "Influence of choice of FLAC and PLAXIS interface models on reinforced soil–structure interactions", *Computers and Geotechnics*, vol. 65, pp. 164–174, 2015.

4

Macroelements for Soil–Structure Interaction

4.1. Introduction

The influence of soil–structure interaction (SSI) on the dynamic response of structures – such as bridges, towers or viaducts – is the object of a significant number of recent studies. In the current practice, SSI is often disregarded, because of the difficulty in modeling it, and also because it is often assumed that the increase in the fundamental period of the soil–structure system (due to the deformability of the soil and foundation) along with the increase in the damping coefficient (due to hysteretic behavior and radiation damping) have beneficial effects on the structural behavior [CIA 95, ELN 96]. Recent works [GAZ 98, MYL 00, JER 04], however, have revealed that stresses in the structure can be amplified in the presence of SSI depending on the characteristics of the loading and of the soil–structure system. Ignoring SSI can therefore lead to a non-conservative assessment of the safety level.

The nonlinear finite element method can be adopted to simulate complex dynamic SSI phenomena. It necessitates, however, discretizing the structure, the foundation and the underlying soil volume, and adopting constitutive models capable of reproducing the soil and the structural material (e.g. concrete and steel) behavior under cyclic/dynamic conditions. Several examples can be found in the literature, see, for example, [ELG 08, JER 09, KOT 13, CHO 19]. The main disadvantage of the approach lies in the necessary dimension of soil volume interacting with the foundation, particularly under seismic conditions, leading to a significant number of degrees of freedom for the soil discretization. The method also requires a detailed geotechnical characterization of the soil deposit and of the structural materials (properties and the

Chapter written by Diana SALCIARINI, Stéphane GRANGE, Claudio TAMAGNINI and Panagiotis KOTRONIS.

exact position of steel reinforcement) and absorbing boundaries to accurately simulate the energy radiation in a deformable medium. The resulting computational cost is prohibitive in everyday engineering practice and often requires massive parallelization numerical tools, especially when strong nonlinearities are present in the structure and in the soil.

On the other hand, simplified tools, such as the Winkler-type models (distributed 1D springs (and dashpots) across the interface between the soil and the footing) should not be preferred in view of all the uncertainties underlying the choice of their parameters [FAR 15]. Their values and distribution vary with frequency, and there is no unique distribution that preserves the global foundation stiffness for all degrees of freedom; in that case, the rocking stiffness is not correctly matched [FAR 15]. The same limitations apply, perhaps even to a larger extent, to pile models developed within the same approach [LI 16]. Winkler-type models require, as input data, a series of soil pressure–displacement (p–y) curves, which are very difficult to select in the absence of data from instrumented lateral pile load tests. For example, Murchison and O'Neill [MUR 84], in a study comparing four proposed procedures for selecting p–y curves with data from field tests, showed that errors in pile head deflection predictions could be as large as 75%. It is also uncertain how the p–y curves are affected by pile head constraints and the relative stiffness of the pile and soil [LI 16].

A significant step forward in the development of efficient, robust and accurate calculation tools for SSI is the so-called *macroelements*, a term coined for the first time by [NOV 91], for the description of the soil–foundation system response by a single constitutive equation that describes the overall behavior of the system. A macroelement can be viewed as an "upscaled" version of classical macroscopic models considering the soil as a continuous (albeit sometimes multiphase) medium where the global behavior of the foundation and of the soil volume interacting with it is "lumped" into a single, integral, constitutive equation that links the evolution of the resultant loads/moments on the foundation to the corresponding displacement/rotation histories [LI 16]. A similar idea in structural engineering to model plastic hinge areas in beam–column joints can be found in [ELA 92] and [DAV 98].

The key element for modeling some of the fundamental features of the global behavior of the soil–foundation system (nonlinearity, irreversibility, dependence on previous load history, coupling between the different degrees of freedom of the system) consists of formulating the constitutive equations of the *macroelement* in incremental form, that is, in the form of evolution laws for the variables of the system. Therefore, it is not surprising that the first applications of the macroelement concept are based on the principles of hardening plasticity [NOV 91] to reproduce the hysteretic response of a shallow soil–foundation system. Macroelements for shallow foundations for monotonic loads are among others proposed by [MON 97, GOT 99, MAR 01, LEP 01, HOU 02, BIE 06] and for cyclic/dynamic loads by [CRÉ 01, CRÉ 02, PRI 03, PRI 06, CHA 08] and [GRA 08a, GRA 08b, GRA 09].

An alternative approach to the theory of elastoplasticity was used by [SAL 09], who derived the constitutive equations of an inelastic macroelement from the basic principles of the generalized hypoplasticity theory [NIE 02]. To extend the basic formulation of the macroelement to the cyclic/dynamic load conditions, the model was enriched by a suitable kinematic internal variable (called the *internal displacement*), adapting the approach proposed by [NIE 97] for continuous media. The hypoplastic macroelement is characterized by the absence of an elastic domain in the space of the generalized stresses and by a continuous variation of the stiffness matrix of the system, depending on the direction of the generalized velocity (*incremental nonlinearity*). Following the work of [SAL 09], macroelements based on hypoplasticity are developed for single vertical piles [LI 16], single inclined piles [LI 18], pile groups [PER 20] and caisson foundations [JIN 19b].

The aim of this chapter is, on the one hand, to show the potentiality offered by the macroelements for the quantitative evaluation of SSI effects and, on the other, to compare the two alternative approaches in the formulation of macroelements (based on elastoplasticity and hypoplasticity). For this purpose, the two macroelements proposed by [GRA 09] and [SAL 09] are used in the numerical simulation of the seismic response of a three-pier viaduct, taken as a benchmark for structures of relevant practical interest.

The chapter is organized as follows. The concept of generalized forces is presented in section 4.2. The general formulation of the incremental constitutive equations, provided by the theories of elastoplasticity and hypoplasticity for the inelastic macroelements, is given in section 4.3. The details of the constitutive equations of the two macroelements are provided in section 4.4. Section 4.5 presents a case study. To compare the two macroelements, particular attention is paid to the calibration of their parameters in section 4.6. The main results of the numerical simulations are discussed and compared in section 4.7. Conclusive remarks are finally summarized in section 4.8.

4.2. The concept of generalized forces: Eurocode 8 recommendations

In the formulation of the two macroelements considered in this chapter, the constitutive equations are written in terms of generalized forces and conjugated generalized displacements. This is motivated by the well-known definition of the bearing capacity of foundations, $N_u = Sq_{max}$, where S is the footing area and q_{max} is the ultimate compression stress of the soil under a vertical centered load (see [DAV 73, MAT 79, PHI 03, RAN 04]), given by

$$q_{max} = \frac{a_s}{2}\gamma B N_\gamma + q_0 N_q + b_s c N_c \qquad [4.1]$$

where q_0 is the vertical effective stress at the bottom of the foundation; N_γ, N_q and N_c are the bearing capacity factors depending on the soil friction angle φ – see, for example, the relations provided by [CAQ 66] and [RAN 04]; γ is the soil unit weight; B is the footing dimension; and c is the soil cohesion. a_s and b_s are the shape factors defined in Table 4.1.

	Circular	Rectangular	Strip ($A \to \infty$)
a_s	0.6	$(1 - 0.2\frac{B}{A})$	1
b_s	1.3	$(1 + 0.2\frac{B}{A})$	1

Table 4.1. *Shape factors for the bearing capacity of circular, rectangular and strip foundations (note: A, B = the length and width of the rectangular footing)*

More recently, several authors provided the ultimate bearing capacity failure conditions by means of suitable failure criteria defined in the space of generalized forces applied to the foundation (see Figure 4.1, where $V = N_{Ed}$ is the vertical force, $H = V_{Ed}$ is the horizontal force and $M = M_{Ed}$ is the moment applied at the center of the foundation).

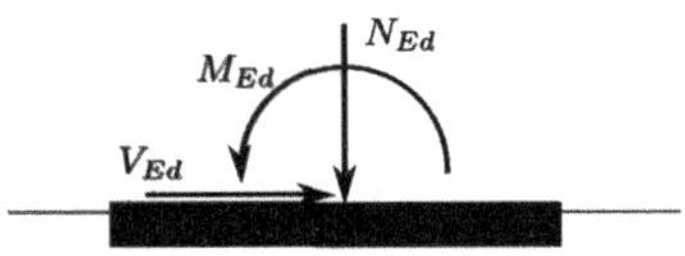

Figure 4.1. *Generalized forces for a shallow foundation*

In the work by [NOV 91], the shape of the failure locus of a shallow foundation, expressing a failure criteria in graphical form, was, for the first time, experimentally defined under a combination of generalized forces for cohesive or frictional soils. [BUT 94, CAS 02, CAS 04, MAR 94, GOT 99] and recently [CHA 07] also provided failure criteria for shallow foundations (see also [LI 14] for pile and [JIN 19a] for caisson foundations). Figure 4.2 shows a comparison between several failure criteria for shallow foundations when projected in different planes.

Chapter 5 of Eurocode 8 [EUR 04], related to SSI problems under seismic loading conditions, provides its own criterion for shallow foundations in terms of generalized forces (equation [4.2]).

$$\frac{\left(1 - e\bar{F}\right)^{c_T}\left(\beta\bar{V}\right)^{c_T}}{\bar{N}^a\left[\left(1 - m\bar{F}^k\right)^{k'} - \bar{N}\right]^b} + \frac{\left(1 - f\bar{F}\right)^{c_M}\left(\gamma\bar{M}\right)^{c_M}}{\bar{N}^c\left[\left(1 - m\bar{F}^k\right)^{k'} - \bar{N}\right]^d} - 1 \leq 0 \qquad [4.2]$$

where the dimensionless variables $\bar{N} = \frac{\gamma_{Rd}N_{Ed}}{N_{\max}}$, $\bar{V} = \frac{\gamma_{Rd}V_{Ed}}{N_{\max}}$ and $\bar{M} = \frac{\gamma_{Rd}M_{Ed}}{BN_{\max}}$ are defined in terms of the vertical design load (N_{Ed}), the design horizontal load

(V_{Ed}), the design moment (M_{Ed}), the maximum vertical load (N_{max}), the soil unit weight (γ) and the footing dimension (B). The parameter $\bar{F}$ represents the horizontal inertial force of the soil, and a, b, c, d, e, f, m, β, γ, c_M and c_T are the parameters depending on the cohesive or frictional character of the soil. Figure 4.3 represents the failure surface in the $M - V$ space, which is described by equation [4.2] for a purely cohesive soil $(\phi = 0)$ and a purely frictional soil $(c = 0)$.

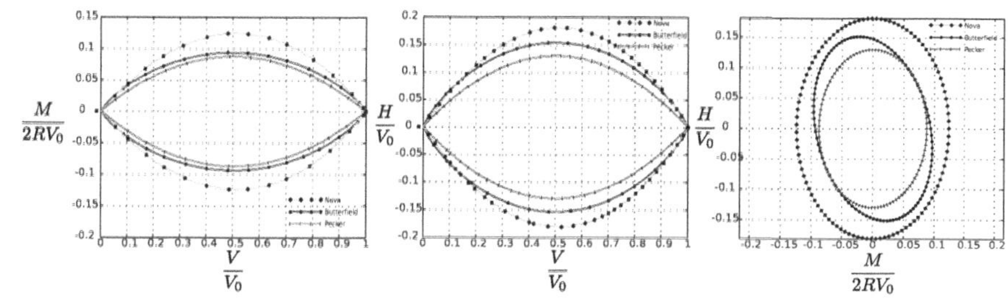

Figure 4.2. *Comparisons between different failure surfaces plotted with dimensionless variables: strip foundations for [NOV 91] and [PEC 97] and circular footings for [BUT 94]. For a color version of this figure, see www.iste.co.uk/grange/soil.zip*

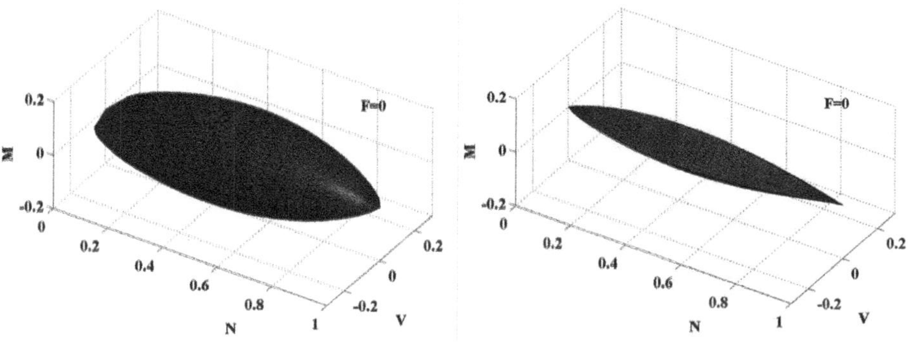

(a) purely cohesive soil (b) purely frictional soil

Figure 4.3. *Representation of the bearing capacity for a shallow foundation from Eurocode 8 (a) for a purely cohesive soil and (b) for a purely frictional soil. For a color version of this figure, see www.iste.co.uk/grange/soil.zip*

Failure criteria are needed to evaluate the resistance of a foundation under a combination of vertical and horizontal forces or moments. However, if displacements have to be estimated, the ultimate failure conditions are not sufficient. For this reason, nonlinear models based, for example, on plasticity mechanisms and failure criteria, are developed to describe the reversible and irreversible load–displacement behavior of the foundation–soil system, under cyclic or dynamic loadings.

4.3. Macroelements for shallow foundations

4.3.1. *Generalities*

The response of the soil–foundation system under given loading conditions is described by means of a single constitutive equation that links the vector of the generalized load applied to the foundation:

$$\boldsymbol{T} := \left\{V\,,\,H_x\,,\,M_y/B_x\,,\,H_y\,,\,M_x/B_y\right\}^T \qquad [4.3]$$

to the *history* of the generalized displacement, collected in the vector

$$\boldsymbol{U} := \left\{U_z\,,\,U_x\,,\,\Theta_y B_x\,,\,U_y\,,\,\Theta_x B_y\right\}^T \qquad [4.4]$$

Here, B_x and B_y represent two characteristic dimensions of the footing, in the horizontal $x-$ and $y-$directions, as shown in Figure 4.4. The torsion moment M_z and the rotation Θ_z are neglected; the extension to the six-dimensional degrees of freedom is discussed in [BIE 06].

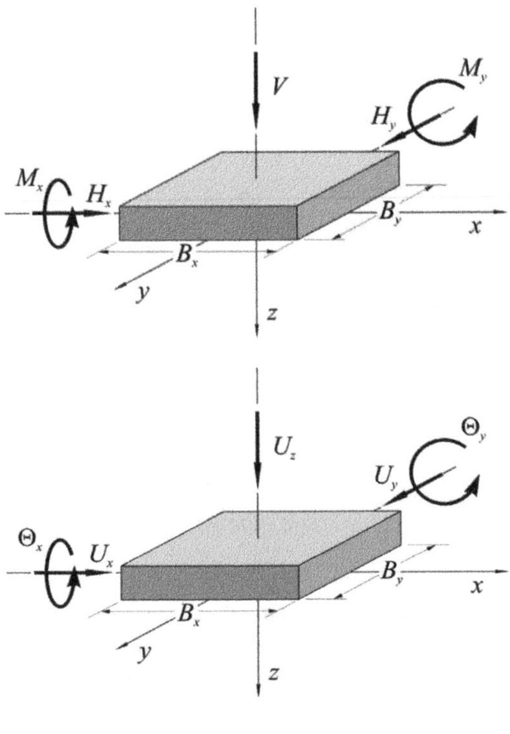

Figure 4.4. *Components of $\boldsymbol{T}$ e $\boldsymbol{U}$: (a) generalized forces and (b) generalized displacements*

To reproduce correctly some important features of the experimentally observed behavior of the foundation–soil system such as nonlinearity, irreversibility and dependence on the past loading history, the constitutive equation for the macroelement must be formulated in rate form, as an *evolution law* of the generalized loads:

$$\dot{\boldsymbol{T}} = \boldsymbol{F}\left(\boldsymbol{T}, \boldsymbol{Q}, \dot{\boldsymbol{U}}\right) \qquad [4.5]$$

starting from a known state, which is defined by $\boldsymbol{T}$ and $\boldsymbol{Q}$. The latter is a vector that collects the internal variable, accounting for the effect of the previous loading history on the system response.

It is worthwhile expressing the generalized forces and the displacements in the following non-dimensional form [GRA 08a, GRA 08b]:

$$\boldsymbol{t} = \left\{v\,, h_x\,, m_y\,, h_y\,, m_x\right\}^T := \frac{1}{V_f}\,\boldsymbol{T} \qquad [4.6]$$

$$\boldsymbol{u} = \left\{u_z\,, u_x\,, \theta_y\,, u_y\,, \theta_x\right\}^T := \frac{\ell}{A}\,\boldsymbol{U} \qquad [4.7]$$

where V_f is the footing bearing capacity under a vertical centered load; $A = B_x B_y$ is the footing area, and $\ell = \sqrt{B_x^2 + B_y^2}$ is a characteristic length accounting for the size of the footing.

With the definitions of [4.6] and [4.7], the constitutive equation for the macroelement can be written in the following non-dimensional form:

$$\dot{\boldsymbol{t}} = \boldsymbol{f}\left(\boldsymbol{t}, \boldsymbol{q}, \dot{\boldsymbol{u}}\right) \qquad [4.8]$$

where $\boldsymbol{q}$ is the non-dimensional internal variable vector.

Since the properties of $\boldsymbol{f}(\boldsymbol{t}, \boldsymbol{q}, \dot{\boldsymbol{u}})$ are selected according to the basic features of observed behavior, $\boldsymbol{f}$ must depend on the loading direction to reproduce an inelastic behavior [KOL 91], and equation [4.8] can have the following alternative form:

$$\dot{\boldsymbol{t}} = \boldsymbol{K}\left(\boldsymbol{t}, \boldsymbol{q}, \boldsymbol{\eta}\right)\dot{\boldsymbol{u}} \qquad [4.9]$$

where $\boldsymbol{K}$ is the tangent stiffness matrix of the system, $\boldsymbol{t}$ is the function of the current loading state, $\boldsymbol{q}$ is the previous loading state and the unit vector:

$$\boldsymbol{\eta} := \frac{\dot{\boldsymbol{u}}}{\|\dot{\boldsymbol{u}}\|}$$

which describes the loading direction.

To construct the function $\boldsymbol{K}$, two different strategies are elaborated hereafter: the theories of elastoplasticity and hypoplasticity.

4.3.2. *Macroelements formulated in the framework of hardening elastoplasticity*

For macroelements developed in the framework of the theory of elastoplasticity [NOV 91, MAR 01, CRÉ 01, GRA 08b], the constitutive equation is built starting from the fundamental assumptions of:

– elastic and plastic decomposition of the generalized velocity:

$$\dot{\boldsymbol{u}} = \dot{\boldsymbol{u}}^e + \dot{\boldsymbol{u}}^p \qquad [4.10]$$

– existence of an elastic law:

$$\dot{\boldsymbol{t}} = \boldsymbol{K}^e(\boldsymbol{t})\,\dot{\boldsymbol{u}}^e \qquad [4.11]$$

where $\boldsymbol{K}^e$ is the elastic stiffness matrix of the soil foundation system;

– existence of a yield function $f_y(\boldsymbol{t}, \boldsymbol{q})$ for each mechanism in the generalized load space;

– existence of a plastic potential function $g(\boldsymbol{t}, \boldsymbol{q})$, evaluated by means of the following flow rule:

$$\dot{\boldsymbol{u}}^p = \dot{\gamma}\frac{\partial g}{\partial \boldsymbol{t}}(\boldsymbol{t}, \boldsymbol{q}) \qquad \dot{\gamma} \geq 0 \qquad [4.12]$$

where $\dot{\gamma}$ is the plastic multiplier;

– existence of a suitable hardening law for the internal variables

$$\dot{\boldsymbol{q}} = \dot{\gamma}\boldsymbol{h}(\boldsymbol{t}, \boldsymbol{q}) \qquad [4.13]$$

– enforcement of Prager's consistency condition.

The resulting constitutive equation in rate form can then be written as

$$\dot{\boldsymbol{t}} = \boldsymbol{K}^{ep}(\boldsymbol{t}, \boldsymbol{q})\,\dot{\boldsymbol{u}} \qquad \boldsymbol{K}^{ep} := \boldsymbol{K}^e - \frac{H(\dot{\gamma})}{K_p}\left\{\boldsymbol{K}^e\left(\frac{\partial g}{\partial \boldsymbol{t}}\right)\left(\frac{\partial f_y}{\partial \boldsymbol{t}}\right)^T \boldsymbol{K}^e\right\} \qquad [4.14]$$

$$\dot{\boldsymbol{q}} = \boldsymbol{H}(\boldsymbol{t}, \boldsymbol{q})\,\dot{\boldsymbol{u}} \qquad \boldsymbol{H} := \frac{H(\dot{\gamma})}{K_p}\boldsymbol{h}\left(\frac{\partial f_y}{\partial \boldsymbol{t}}\right)^T \boldsymbol{K}^e \qquad [4.15]$$

where $K_p > 0$ is a scalar quantity given by

$$K_p := \left(\frac{\partial f_y}{\partial \boldsymbol{T}}\right)^T \boldsymbol{K}^e \frac{\partial g}{\partial \boldsymbol{T}} - \left(\frac{\partial f_y}{\partial \boldsymbol{T}}\right)^T \boldsymbol{H}_q \qquad [4.16]$$

and $H(\dot{\gamma})$ is the Heaviside step function, equal to one if the plastic multiplier:

$$\dot{\gamma} := \frac{1}{K_p}\left(\frac{\partial f}{\partial t}\right)^T \boldsymbol{K}^e \dot{\boldsymbol{u}}$$

is positive and zero otherwise.

4.3.3. *Macroelements formulated in the framework of hypoplasticity*

In the approach based on the theory of hypoplasticity, the constitutive function $\boldsymbol{f}$ is constructed starting from the following general expression:

$$\dot{\boldsymbol{t}} = \boldsymbol{L}\left(\dot{\boldsymbol{u}} - Y\boldsymbol{m}\,\|\dot{\boldsymbol{u}}\|\right) \qquad [4.17]$$

where the matrix (5×5) $\boldsymbol{L}(\boldsymbol{t}, \boldsymbol{q})$, the unit vector $\boldsymbol{m}(\boldsymbol{t}, \boldsymbol{q})$ and the scalar quantity $Y(\boldsymbol{t})$ (varying between 0 and 1) are constitutive equations, built from observed behavior of the system, as better described in the following.

The application of the Euler theorem to the second term of equation [4.17] can be written as

$$\dot{\boldsymbol{t}} = \boldsymbol{K}^{hp}\dot{\boldsymbol{u}} \qquad \boldsymbol{K}^{hp} := \boldsymbol{L} - Y\boldsymbol{m}\boldsymbol{\eta}^T \qquad [4.18]$$

where $\boldsymbol{K}^{hp}(\boldsymbol{t}, \boldsymbol{q}, \boldsymbol{\eta})$ is the tangent stiffness matrix of the system. Note that, unlike the theory of hardening plasticity, where the stiffness matrix can assume only two distinct values (elastic and elastoplastic), in the theory of hypoplasticity, the tangent stiffness of the system is a continuous function of $\boldsymbol{\eta}$. This property is named *incremental nonlinearity* by [DAR 78].

The physical meaning of the three constitutive equations $\boldsymbol{L}$, $\boldsymbol{m}$ and Y can be explained assuming that the scalar function Y (*loading function*) has values close to zero within the failure surface $\mathcal{S}_f$, and monotonically increases when the distance from $\mathcal{S}_f$ to the current loading state decreases, up to reach the unit value for $\boldsymbol{t} = \boldsymbol{t}_f \in \mathcal{S}_f$.

When the current state is that $Y \to 0$, equation [4.17] reduces to:

$$\dot{\boldsymbol{t}} \simeq \boldsymbol{L}\dot{\boldsymbol{u}}$$

and the response of the system is approximately linear. In this condition, the matrix $\boldsymbol{L}$ can be assimilated to the elastic stiffness matrix $\boldsymbol{K}^e$ of the theory of elastoplasticity.

If the soil–foundation system is subjected to displacements having constant velocity module and direction ($\dot{\boldsymbol{u}} = \|\dot{\boldsymbol{u}}\|\,\boldsymbol{\eta} = C\boldsymbol{\eta}$ constant), in failure conditions ($Y = 1$), equation [4.17] becomes:

$$\dot{\boldsymbol{t}} = C\boldsymbol{L}$$

Hence, the vector $\boldsymbol{m}$ represents the direction at failure, associated with the loading state $\boldsymbol{t}_f$ that is asymptotically reached by the system.

To correctly describe the response of the soil–foundation system under cyclic loads, [SAL 09] introduced the following modification to the constitutive equation [4.18], starting from the extension proposed by [NIE 97] for the hypoplastic constitutive model of [WOL 96].

Let us introduce an internal variable $\boldsymbol{\delta}$, called the *internal displacement*, with the following evolution law:

$$\dot{\boldsymbol{\delta}} = \begin{cases} \left(\boldsymbol{I} - \rho^{\beta_r}\boldsymbol{\eta}_\delta\boldsymbol{\eta}_\delta^T\right)\dot{\boldsymbol{u}} & (\boldsymbol{\eta}_\delta \cdot \boldsymbol{\eta} > 0) \\ \dot{\boldsymbol{u}} & (\boldsymbol{\eta}_\delta \cdot \boldsymbol{\eta} \leq 0) \end{cases} \qquad [4.19]$$

where $\boldsymbol{\eta}_\delta := \boldsymbol{\delta}/\left\|\boldsymbol{\delta}\right\|$ is the unit vector, which provides the direction of the internal displacement; $\rho \in [0,1]$ is a normalized measure of the module of $\boldsymbol{\delta}$; β_r is a model constant, and $\boldsymbol{I}$ is the identity matrix (5×5). The evolution law [4.19] is constructed to have internal displacements that tend to be tangent to the path of the generalized displacements (Figure 4.5).

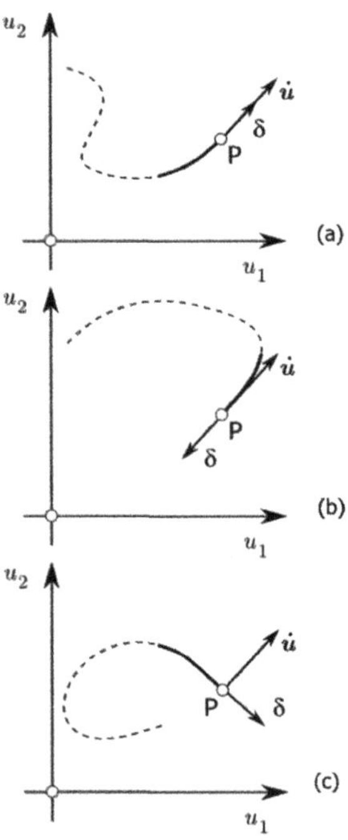

Figure 4.5. *Effect of the displacement history on the system response, for a given loading state, varying the direction of the generalized velocity* $\dot{\boldsymbol{u}}$

When including the internal displacement $\boldsymbol{\delta}$, the constitutive equation of the macroelement [4.18] modifies as follows ([SAL 09]):

$$\dot{\boldsymbol{t}} = \boldsymbol{K}^{hp}(\boldsymbol{t}, \boldsymbol{q}, \boldsymbol{\delta})\,\dot{\boldsymbol{u}} \quad [4.20]$$

where:

$$\boldsymbol{K}^{hp}(\boldsymbol{t}, \boldsymbol{q}, \boldsymbol{\delta}) := \left[\rho^{\chi} m_T + (1 - \rho^{\chi})\, m_R\right] \boldsymbol{L}(\boldsymbol{t}, \boldsymbol{q}) + \widetilde{\boldsymbol{K}}(\boldsymbol{t}, \boldsymbol{q}, \boldsymbol{\delta}) \quad [4.21]$$

and

$$\widetilde{\boldsymbol{K}}(\boldsymbol{t}, \boldsymbol{q}, \boldsymbol{\delta}) := \begin{cases} \rho^{\chi}(1 - m_T)(\boldsymbol{L}\boldsymbol{\eta}_{\delta})\,\boldsymbol{\eta}_{\delta}^{T} - \rho^{\chi} Y \boldsymbol{m} \boldsymbol{\eta}_{\delta}{}^{T} & (\boldsymbol{\eta}_{\delta} \cdot \boldsymbol{\eta} > 0) \\ \rho^{\chi}(m_R - m_T)(\boldsymbol{L}\boldsymbol{\eta}_{\delta})\,\boldsymbol{\eta}_{\delta}^{T} & (\boldsymbol{\eta}_{\delta} \cdot \boldsymbol{\eta} \leq 0) \end{cases} \quad [4.22]$$

with m_R, m_T and χ being model constants.

Due to the evolution law [4.19], the internal displacement $\boldsymbol{\delta}$ takes into account the displacement history which occurred during the previous loading stages. Varying the generalized velocity direction with respect to the internal displacement direction, the constitutive equation [4.21] can assume the following forms:

a) For proportional loading, $\rho \to 1$ and $\boldsymbol{\eta}_{\delta} \to \boldsymbol{\eta}$ (Figure 4.5a); in these conditions, from equation [4.21] and equation [4.22]:

$$\dot{\boldsymbol{t}} = \boldsymbol{L}\left(\dot{\boldsymbol{u}} - Y\boldsymbol{m}\,\|\dot{\boldsymbol{u}}\|\right)$$

i.e. equation [4.17] of the original hypoplastic model.

b) For a loading reversal, starting from a hypoplastic state (i.e. $\boldsymbol{\eta} = -\boldsymbol{\eta}_{\delta}$ and $\rho = 1$), we have:

$$\boldsymbol{K}^{hp} = m_R \boldsymbol{L} \quad [4.23]$$

and the system response is locally (hypo)elastic.

c) For a loading path orthogonal to the internal displacement direction, starting from a hypoplastic state (i.e. $\boldsymbol{\eta} \cdot \boldsymbol{\eta}_{\delta} = 0$ and $\rho = 1$), we have:

$$\boldsymbol{K}^{hp} = m_T \boldsymbol{L} \quad [4.24]$$

and the system response is again locally (hypo)elastic, but characterized by a different stiffness (typically $m_T \leq m_R$).

In all the intermediate cases, the expression of the tangent stiffness of the system is obtained by interpolation.

4.4. The considered macroelements

Two macroelements for shallow foundations are considered and compared: the elastoplastic macroelement with isotropic hardening proposed by [GRA 08a, GRA 08b, GRA 09] for the SSI analysis in seismic conditions, and the hypoplastic macroelement by [SAL 09]. In the following, a brief description of the relevant aspects of the two formulations is presented; complete details are given in the cited references.

4.4.1. *The elastoplastic macroelement*

The elastoplastic macroelement by [GRA 09] considers the following decomposition of the generalized velocity:

$$\dot{\boldsymbol{u}} = \dot{\boldsymbol{u}}^e + \dot{\boldsymbol{u}}^p + \dot{\boldsymbol{u}}^{up} \qquad [4.25]$$

The velocities $\dot{\boldsymbol{u}}^p$ and $\dot{\boldsymbol{u}}^{up}$ refer to two different mechanisms of plastic deformation: the first mechanism is related to the irreversible deformations that develop in the soil when the applied loads reach the failure conditions of the foundation; the second mechanism is due to the partialization of the contact area between soil and foundation (phenomenon defined as "uplift" by [GRA 09]). In the following, this second mechanism is not considered.

The elastic component of the generalized velocity is provided by the elastic law [4.11], where the stiffness matrix is given by

$$\boldsymbol{K}^e = \text{diag}\,\{k_{vv}\,; k_{hx}\,; k_{my}\,; k_{hy}\,; k_{mx}\} \qquad [4.26]$$

with

$$k_\alpha = \frac{\sqrt{A}}{\ell}\,\frac{G}{q_f}\,\frac{1}{(1-\nu)}\beta_\alpha \qquad [4.27]$$

and $\alpha \in \{vv\,; hx\,; my\,; hy\,; mx\}$. In equation [4.26], G and ν are the shear modulus and the Poisson ratio of the soil, under the hypothesis of linear elastic, homogeneous and isotropic half-space; $q_f := V_f/A$ is the ultimate bearing capacity of the foundation under a vertical centered load, and the non-dimensional coefficients β_α are given by [PHI 03] as function of the ratio between the footing size B_x/B_y.

The elastic domain of the macroelement is defined starting from the following yield function:

$$f_y(\boldsymbol{t}, \boldsymbol{\tau}, \gamma, \rho) = \left\{\frac{h_x}{\rho a v^c(\gamma - v)^d} - \frac{\tau_{hx}}{\rho}\right\}^2 + \left\{\frac{m_y}{\rho b v^e(\gamma - v)^f} - \frac{\tau_{my}}{\rho}\right\}^2$$
$$+ \left\{\frac{h_y}{\rho a v^c(\gamma - v)^d} - \frac{\tau_{hy}}{\rho}\right\}^2 + \left\{\frac{m_x}{\rho b v^e(\gamma - v)^f} - \frac{\tau_{mx}}{\rho}\right\}^2 - 1 = 0 \qquad [4.28]$$

where a, b, c, d, e and f are the model constants, while $\boldsymbol{\tau}$, ρ and γ are the internal variables controlling the position, size and shape of the failure surface.

The failure surface can be obtained by equation [4.28], putting $\boldsymbol{\tau} = \mathbf{0}$ and $\rho = \gamma = 1$:

$$f_c(\boldsymbol{t}) = \left\{\frac{h_x}{av^c(1-v)^d}\right\}^2 + \left\{\frac{m_y}{bv^e(1-v)^f}\right\}^2 + \left\{\frac{h_y}{av^c(1-v)^d}\right\}^2 + \left\{\frac{m_x}{bv^e(1-v)^f}\right\}^2 - 1 = 0 \quad [4.29]$$

A plot of a possible yield surface and the failure surface is given in Figure 4.6.

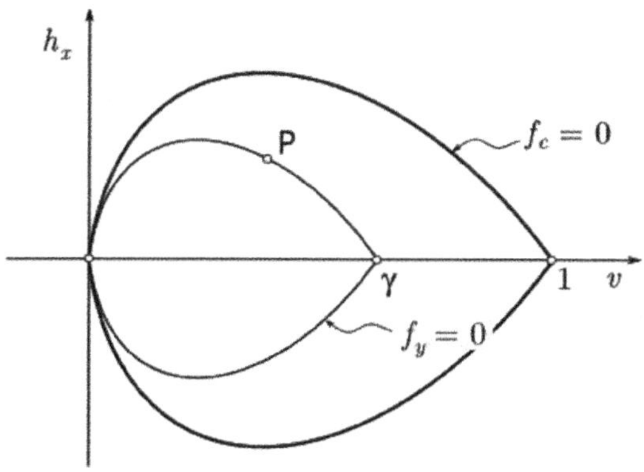

Figure 4.6. *Elastoplastic model by [GRA 09]: evolution of the yield surfaces with the plastic deformation and failure surface*

In plastic conditions ($f = 0$ and $\dot{f} = 0$), the plastic component of the generalized velocity is given by equation [4.12], with the following plastic potential function:

$$g(\boldsymbol{t}, \boldsymbol{\tau}, \gamma, \rho, v_0) = \left\{\frac{h_x(\kappa\gamma + v_0)^d}{\rho a v_0^c(\gamma - v)^d(\kappa\gamma + v)^d} - \frac{\tau_{hx}}{\rho}\right\}^2 + \left\{\frac{m_y(\xi\gamma + v_0)^f}{\rho b v_0^e(\gamma - v)^f(\xi\gamma + v)^f} - \frac{\tau_{my}}{\rho}\right\}^2 + \left\{\frac{h_y(\kappa\gamma + v_0)^d}{\rho a v_0^c(\gamma - v)^d(\kappa\gamma + v)^d} - \frac{\tau_{hy}}{\rho}\right\}^2 + \left\{\frac{m_x(\xi\gamma + v_0)^f}{\rho b v_0^e(\gamma - v)^f(\xi\gamma + v)^f} - \frac{\tau_{mx}}{\rho}\right\}^2 - 1 = 0 \quad [4.30]$$

where κ and ξ are the two model constants and v_0 is the vertical force, normalized with respect to the current state, see [GRA 08a, GRA 08b] for further details. The internal variables $\boldsymbol{\tau}$, γ and ρ control, respectively, the position of the center of the yield surface, its size in the v-direction and its shape. Their evolution laws equation [4.13] derive from the experimental observations by [GOT 99] and from the results of finite element numerical simulations by [CRÉ 01].

The calibration of the elastoplastic macroelement requires the determination of 16 constants described in Table 4.2. The constants B_x and B_y define the geometry of the foundation; k_{vv}, k_{hx}, k_{my}, k_{hy} and k_{my} are the components of the normalized elastic stiffness matrix; a, b, c, d, e and f control the shape of the yield surface; ξ and κ control the direction of the plastic flow in the (h_x, v)-plane, (h_y, v)-plane, (m_x, v)-plane and (m_y, v)-plane, while V_f is the footing bearing capacity under centered vertical load.

Constant	Unit	Description
B_x	[L]	footing size
B_y	[L]	
k_{vv}	[–]	elastic stiffness matrix components
k_{hx}	[–]	
k_{my}	[–]	
k_{hy}	[–]	
k_{mx}	[–]	
a	[–]	yield function parameters
b	[–]	
c	[–]	
d	[–]	
e	[–]	
f	[–]	
ξ	[–]	plastic potential function parameters
κ	[–]	
V_f	[F]	bearing capacity under vertical centered load

Table 4.2. *Description of the constants of the elastoplastic macroelement*

4.4.2. *The hypoplastic macroelement*

In the formulation of the hypoplastic macroelement, [SAL 09] relies on the model by [NOV 91] to define both the loading function Y and the direction of the plastic flow $\boldsymbol{m}$.

In the space of non-dimensional loads, the failure surface proposed by [NOV 91] can be generalized as follows:

$$f_c(\boldsymbol{t}) = \left\{\frac{h_x}{\mu}\right\}^2 + \left\{\frac{m_x}{\psi}\right\}^2 + \left\{\frac{h_y}{\mu}\right\}^2 + \left\{\frac{m_y}{\psi}\right\}^2 - v^2(1-v)^{2\beta} = 0 \quad [4.31]$$

where μ, ψ and β are the model constants. For any "stable" loading state within the failure domain, a surface that is omothetic to the failure surface can be defined as

$$f^*(\boldsymbol{t},\gamma) = \left\{\frac{h_x}{\mu}\right\}^2 + \left\{\frac{m_x}{\psi}\right\}^2 + \left\{\frac{h_y}{\mu}\right\}^2 + \left\{\frac{m_y}{\psi}\right\}^2 - v^2\left(1-\frac{v}{\gamma}\right)^{2\beta} = 0 \quad [4.32]$$

passing for $\boldsymbol{t}$ and intercepting the v-axis at $v = \xi$ (Figure 4.7). When the loading state varies, the parameter γ varies between $(0,1]$ and reaches the value of 1 when the loading state is on the failure surface. Owing by this, the following loading function has been adopted:

$$Y(\boldsymbol{t}) = \{\gamma(\boldsymbol{t})\}^{\kappa} \quad [4.33]$$

where κ is a model constant.

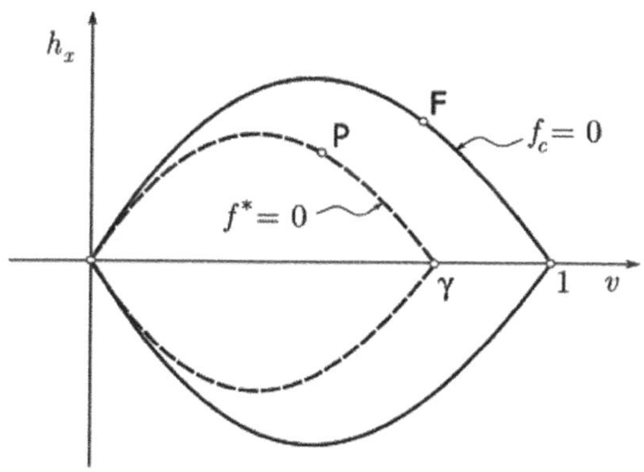

Figure 4.7. *Definition of the surface $f = 0$ and γ*

The plastic flow direction is evaluated as

$$\boldsymbol{m}(\boldsymbol{t}) = \frac{1}{\|\partial g/\partial \boldsymbol{t}\|}\frac{\partial g}{\partial \boldsymbol{t}} \quad [4.34]$$

starting from the generalization of the plastic potential function defined by the model of [NOV 91]:

$$g(\boldsymbol{t},\gamma) = \left\{\frac{h_x}{\lambda_h \mu}\right\}^2 + \left\{\frac{m_x}{\lambda_m \psi}\right\}^2 + \left\{\frac{h_y}{\lambda_h \mu}\right\}^2 + \left\{\frac{m_y}{\lambda_m \psi}\right\}^2 - v^2\left(1 - \frac{v}{\gamma_g}\right)^{2\beta} \quad [4.35]$$

In equation [4.34], λ_h and λ_m are the model constants, while γ_g is a dummy variable, obtained considering the condition $g(\boldsymbol{t},\gamma_g) = 0$ at the current loading state. The geometrical meaning of γ_g is given in Figure 4.8.

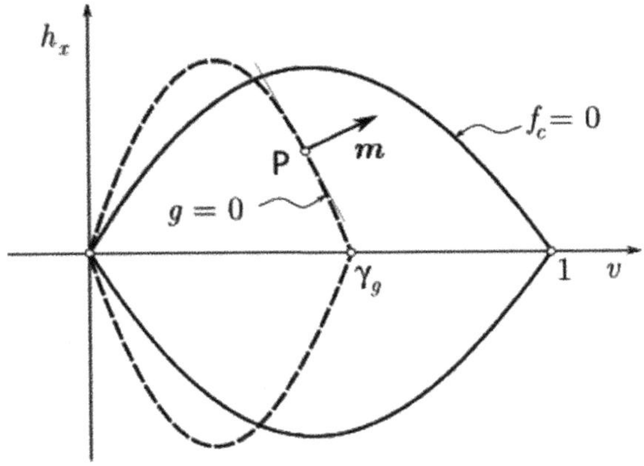

Figure 4.8. *Definition of the surface $g = 0$ and γ_g*

Considering that, after a full reversal of the loading path, the hypoplastic macroelement responds similarly to a linear elastic system, the $\boldsymbol{L}$ matrix of the hypoplastic model can be written as

$$\boldsymbol{L} = \frac{1}{m_R}\boldsymbol{K}^e = \frac{1}{m_R}\,\mathrm{diag}\,\{k_{vv}\,;k_{hx}\,;k_{my}\,;k_{hy}\,;k_{mx}\} \quad [4.36]$$

where k_α are provided by equation [4.27].

To complete the hypoplastic macroelement description, the normalized internal displacement ρ, appearing in equation [4.21] and equation [4.22], has to be defined. The following expression has been adopted by [SAL 09]:

$$\rho = \frac{\|\delta\|_W}{R} := \frac{\sqrt{\delta^T \boldsymbol{W} \delta}}{R} \quad [4.37]$$

where:

$$\boldsymbol{W} := \mathrm{diag}\,\{1\,;w_h\,;w_m\,;w_h\,;w_m\} \quad [4.38]$$

is a diagonal matrix of weights, accounting for the geometry of the failure surface, while R represents the size region within the displacement space where the stiffness of the system is high and the behavior is quasilinear ("elastic nucleus"), see [SAL 09] for further details.

The hypoplastic macroelement is characterized by 19 constants, which are described in Table 4.3. As in the elastoplastic model, B_x and B_y define the footing size, while k_{vv}, k_{hx}, k_{my}, k_{hy} and k_{my} are the components of the normalized elastic stiffness matrix. μ, ψ and β control the shape of the failure domain in the generalized loading space. In particular, μ can be interpreted as the friction coefficient at the soil–foundation interface, while ψ controls the failure condition when vertical eccentric loads are present [NOV 91]. λ_h and λ_m control the plastic flow direction along the planes (h_x, v), (h_y, v), (m_x, v) and (m_y, v), while κ controls the degree of nonlinearity of the system response. Finally, m_R, m_T, R, β_r and χ control the evolution of the stiffness of the system at varying of the loading path direction and history of previous loading conditions.

Constant	Unit	Description
B_x	[L]	footing size
B_y	[L]	
k_{vv}	[–]	components of the stiffness matrix $m_R\boldsymbol{L}$
k_{hx}	[–]	
k_{my}	[–]	
k_{hy}	[–]	
k_{mx}	[–]	
μ	[–]	parameters of the function $F(\boldsymbol{t})$
ψ	[–]	
β	[–]	
λ_h	[–]	parameters of the function $g(\boldsymbol{t})$
λ_m	[–]	
κ	[–]	parameter of the loading function Y
m_R	[–]	stiffness coefficient for $\boldsymbol{\eta} \cdot \boldsymbol{\eta}_\delta = -1$
m_T	[–]	stiffness coefficient for $\boldsymbol{\eta} \cdot \boldsymbol{\eta}_\delta = 0$
R	[L]	size of the zone at high stiffness in the generalized displacement space
β_r	[–]	coefficient of the evolution law for $\boldsymbol{\delta}$
χ	[–]	coefficient of the interpolation law for $\boldsymbol{K}^{hp}$
V_f	[F]	bearing capacity under centered vertical load

Table 4.3. *Description of the constants of the hypoplastic macroelement*

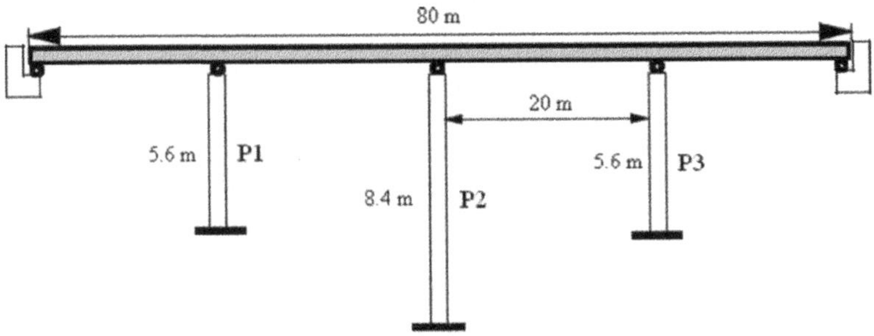

Figure 4.9. *Viaduct scheme (out of scale; the dimensions reported refer to the small-scale prototype)*

4.5. Case study: seismic response of a reinforced concrete viaduct

The potentialities offered by the two macroelements considered in the analysis of SSI phenomena are hereafter evaluated by simulating the seismic response of a four-span viaduct founded on shallow foundations, resting on medium-dense sands. In particular, the following three cases have been considered:

– soil–foundation system modeled by means of the elastoplastic macroelement by [GRA 09] (case "MEP");

– soil–foundation system modeled by means of the hypoplastic macroelement by [SAL 09] (case "MHP");

– foundation on infinitely rigid soil (case "RIG" – no SSI).

The goal of this simulation program is twofold: on the one hand, quantifying the influence of the macroelement mathematical formulation on the prediction of the global system performance by comparing MEP and MHP case solutions; on the other hand, evaluating the effect of SSI on the distribution of structural stresses and displacements by comparing the MEP and MHP case solutions with the one obtained from the RIG case (in the absence of interaction).

4.5.1. *Features of the viaduct*

The considered structure is a four-span viaduct, with a pre-compressed reinforced concrete deck and reinforced concrete piers with rectangular hollow sections (Figure 4.9). The viaduct has been analyzed at the ELSA laboratory (JRC Ispra) using small-scale physical models (1÷2.5) for the three piers and a finite element model for the deck [PIN 96].

The geometric characteristics of the piers and the beam sections are provided in Table 4.4. In the following, the viaduct is simulated by considering a length of the deck of 80 m, a height of the lateral piers (P1 and P3) of 5.6 m, and a height of the lateral pier (P2) of 8.4 m. The piers are arranged symmetrically with respect to the center of the deck, which is considered bounded with a simple roller constraint at the head of the piers.

	A (m^2)	I_x (m^4)	I_y (m^4)	I_z (m^4)	J (m^4)
Deck	1.11	0.13	–	2.26	2.39
Piers	0.66	0.056	0.19	–	0.20

Table 4.4. *Geometric characteristics of the structural elements of the viaduct*

4.5.2. The finite element model of the viaduct and its foundations

For the numerical simulations, the finite element Matlab toolbox FEDEASLab [FIL 04] has been used. The finite element model of the structure is shown in Figure 4.10.

To simulate the behavior of the viaduct piers, subjected to bending and shear stresses, nonlinear Timoshenko beam elements with multi-fiber formulation [KOT 05] have been adopted. The mechanical behavior of the concrete fibers has been described using the damage model by [LAB 91], while a modified version of the model [MEN 73] has been adopted for the steel fibers of the reinforcing cage. For the deck, elastic linear beam elements have been used.

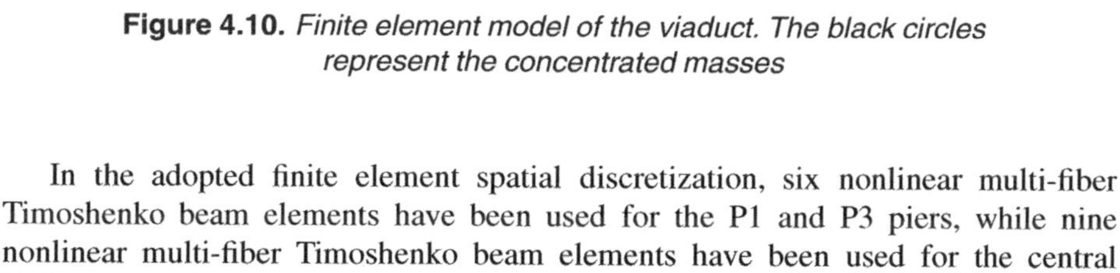

Figure 4.10. *Finite element model of the viaduct. The black circles represent the concentrated masses*

In the adopted finite element spatial discretization, six nonlinear multi-fiber Timoshenko beam elements have been used for the P1 and P3 piers, while nine nonlinear multi-fiber Timoshenko beam elements have been used for the central highest pier. The discretization has been refined at the base of the piers where a higher concentration of stresses is expected. Some details of the multi-fiber discretization of the sections of the nonlinear Timoshenko beams are illustrated in Figure 4.11. For each section, 40 concrete fibers and 80 steel fibers, representing the reinforcing bars in their actual position, have been used.

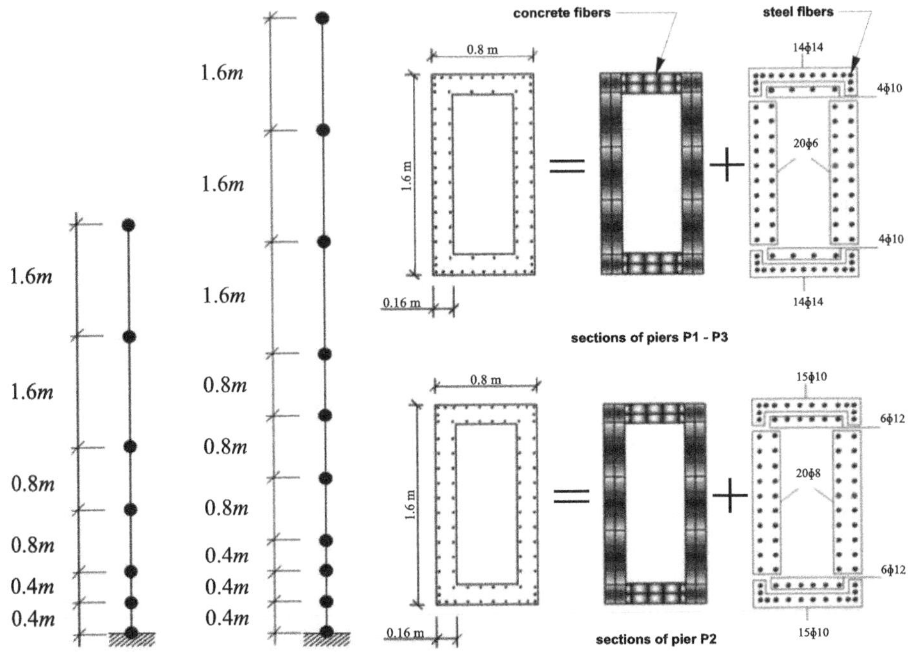

Figure 4.11. *Details of the pier discretization and the different cross-sections*

The inertial characteristics of the structural elements have been simulated through masses concentrated at the element nodes, as shown in Figure 4.10. The values of the constants that characterize the materials of the different structural elements used in the simulations are provided in [GRA 08a].

The two macroelements used to describe the soil–foundation system in MEP and MHP analyses have been implemented in FEDEASLab using different integration algorithms, according to the different nature of their constitutive equations. For the elastoplastic macroelement by [GRA 09], a *predictor–corrector* strategy has been adopted, based on the use of the implicit backward Euler algorithm for the projection phase on the yield surface [GRA 08a]. For the hypoplastic macroelement by [SAL 09] – which does not have a yield surface to impose the consistency condition – an explicit Runge–Kutta–Fehlberg algorithm of third order with adaptive step and error check has been adopted [STO 93].

In the formulation of the two macroelements, the introduction of dissipative elements capable of reproducing the damping phenomenon by radiation has not been provided. This introduces a simplification that may have an effect on the computed response of the system, but this is accepted considering the observations by [JER 04]:

– radiation damping shows its effects primarily at high frequencies and low values of the mechanical damping for hysteresis;

– if the foundation undergoes a partial detachment from the soil, the radiation damping is greatly reduced;

– the simplified approaches used, for example, by [SPY 92] or [ZHA 02], which use dampers with frequency-independent response, tend to produce a generalized reduction of stresses in the structure.

In addition, studies conducted by Bielak *et al.* [BIE 03, YOS 03] indicate that, in some cases, radiation damping can be ignored without significant variations in the structure response.

4.5.3. *Seismic input*

The seismic input adopted in the finite element simulations is shown in Figure 4.12. This is an artificial accelerogram, applied along the x-direction, compatible with the response spectrum provided by Eurocode 8 for a Class B soil and a damping of 5%, characterized by a peak acceleration of 0.35 g. To comply with the conditions of similarity in the 1÷2.5 scale model used in the simulations, the accelerations have been multiplied by 2.5, while the time scale was divided by the same coefficient. The same seismic input has been applied to the abutment of the bridge and to the foundations of the three piers, neglecting differences that may be due to topographic amplification effects.

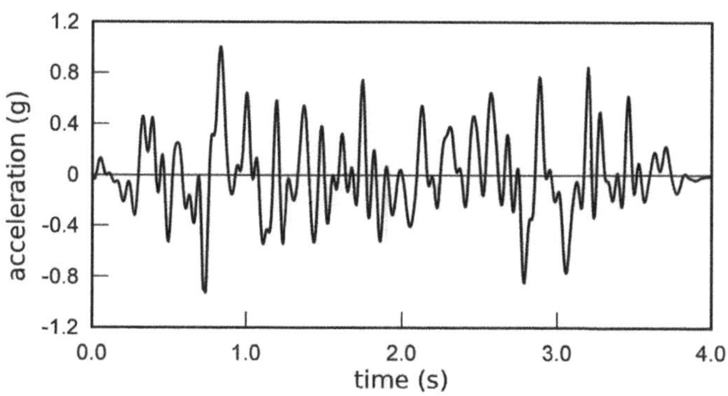

Figure 4.12. *Scaled accelerogram applied to the foundations and abutment of the bridge*

4.6. Calibration of the macroelements

The width B_x and the length B_y of the foundations have been assumed equal to 2.1 m and 4.1 m, respectively. The procedures for determining the remaining 14 constants of the elastoplastic macroelement are described in detail in [GRA 08a]. For the purposes of this chapter, the values reported in Table 4.5 have been adopted, as they can be considered appropriate for foundations on medium-dense sands.

k_v (–)	k_{hx} (–)	k_{my} (–)	k_{hy} (–)	k_{mx} (–)	ξ (–)	κ (–)
84.85	69.42	30.87	69.42	30.87	0.0	0.0

a (–)	b (–)	c (–)	d (–)	e (–)	f (–)	V_f (MN)
0.48	0.33	1.00	0.95	1.00	0.95	11.26

Table 4.5. *Constants of the elastoplastic macroelement adopted in the numerical simulations*

To compare the response provided by the two macroelements, a calibration of the MHP has been conducted on the following stress paths:

– phase a): centered vertical load V monotonously increasing up to 1200 kN; phase b): horizontal shift U_x monotonously increasing up to 0.02 m;

– phase a): centered vertical load V monotonously increasing up to 1200 kN; phase b): rotation Θ_y monotonously increasing up to 0.004 rad;

– phase a): centered vertical load V monotonously increasing up to 1200 kN; phase b): cyclic horizontal displacement:

$$U_x(t) = U_{x,\text{ref}}\hat{\lambda}(t) \qquad\qquad 0 \leq t \leq T$$

applied at a distance h = 0.775 m from the baricenter of the foundation. In the previous expression, $U_{x,\text{ref}} = 0.001$ m, while the load multiplier $\hat{\lambda}(t)$ is represented in Figure 4.13 as a function of the ratio t/T.

The constants that characterize the foundation size and its pseudoelastic behavior, as well as the vertical load in ultimate conditions V_f, have the same meaning of the corresponding constants of the elastoplastic model, and can therefore be assumed identical to those found in Table 4.5.

By comparing the expressions provided by equations equation [4.29] and [4.31], observing that for the elastoplastic macroelement, we assumed $c = e = 1$ and $d = f$, and choosing $\mu = a$, $\psi = b$ and $\beta = d$, the collapse domains of the two models are coincident.

The numerical values of the remaining constants (λ_h, λ_m, κ, m_R, m_T, R, β_r and χ) have been determined by attempts, comparing the response of the two models (MHP, MEP) along the loading paths (1)–(3). This comparison is graphically illustrated in Figures 4.14–4.19. The final result of the calibration procedure of the hypoplastic macroelement is summarized in Table 4.6.

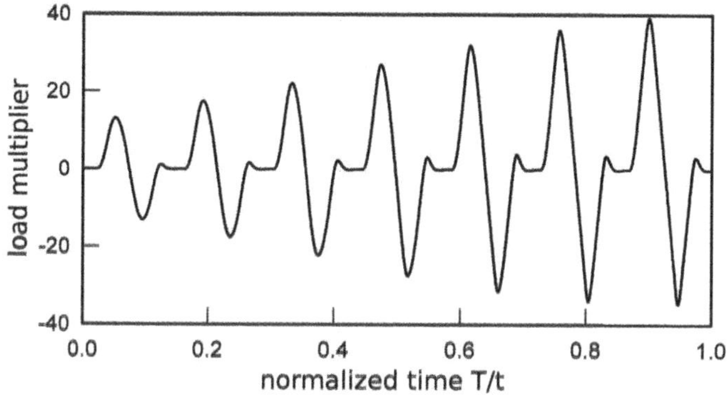

Figure 4.13. *Load multiplier adopted in the cyclic calibration test no. 3*

μ (–)	ψ (–)	β (–)	λ_h (–)	λ_m (–)	κ (–)
0.48	0.33	0.95	1.75	1.50	0.25

m_R (–)	m_T (–)	R (mm)	β_r (–)	χ (–)	V_f (MN)
1.1	1.05	5.0	1.0	1.5	11.26

Table 4.6. *Constants of the hypoplastic macroelement adopted in the numerical simulations*

The figures show that the response of the two models in the three calibration tests is generally very similar, even for cyclic load paths with variable amplitude, such as the one used in test no. 3. The only significant difference concerns the evolution of the vertical displacement calculated during the cyclic phase of this last test, qualitatively similar but quantitatively different (Figure 4.19). Indeed, the maximum settlement of the hypoplastic macroelement at the end of the test is greater than that of the elastoplastic macroelement, of about 130%.

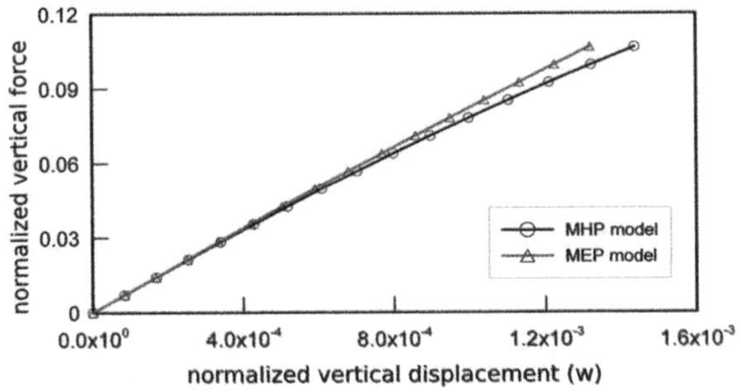

Figure 4.14. *Calibration test no. 1 – phase a): normalized vertical force v versus normalized vertical displacement w. For a color version of this figure, see www.iste.co.uk/grange/soil.zip*

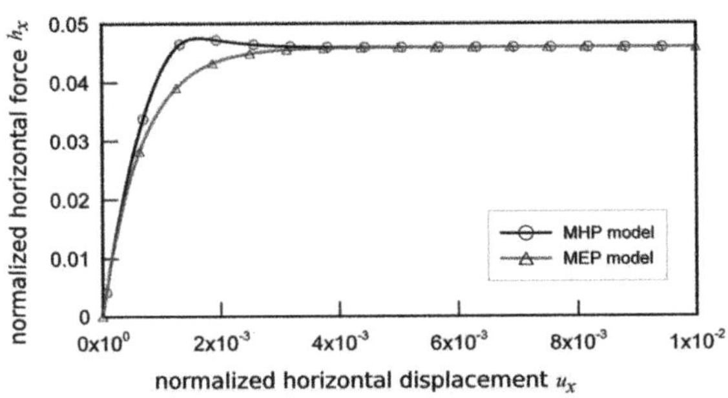

Figure 4.15. *Calibration test no. 1 – phase a): normalized horizontal force h_x versus normalized horizontal displacement u_x. For a color version of this figure, see www.iste.co.uk/grange/soil.zip*

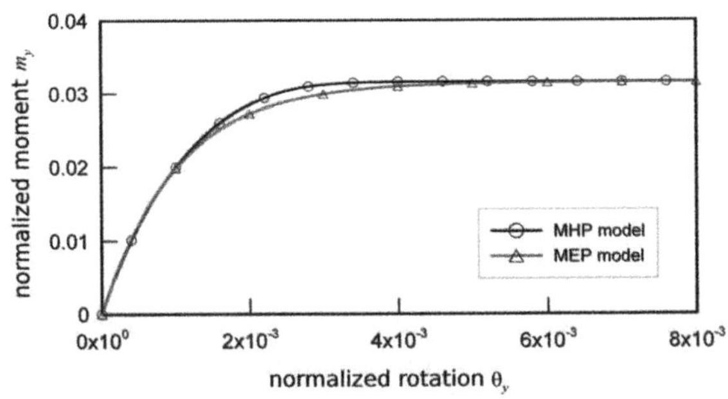

Figure 4.16. *Calibration test no. 2 – phase b): normalized moment m_y versus normalized rotation θ_y. For a color version of this figure, see www.iste.co.uk/grange/soil.zip*

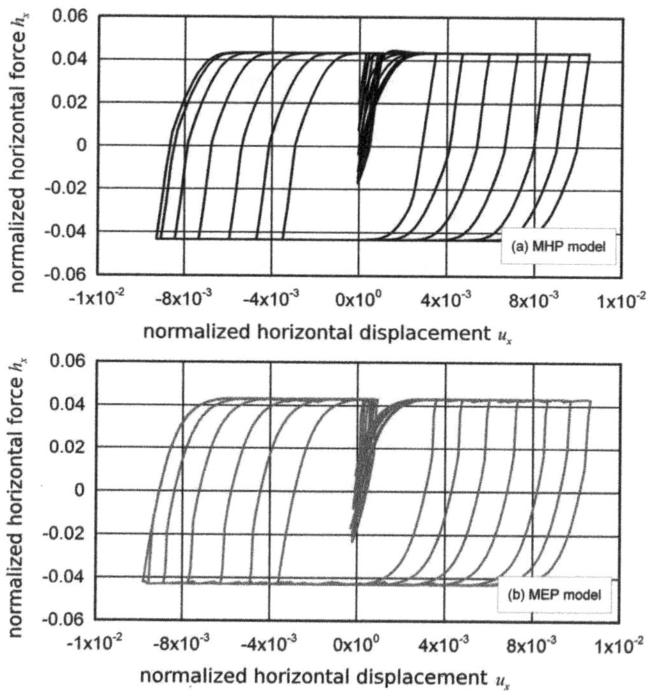

Figure 4.17. *Calibration test no. 3 – phase b): normalized horizontal force h_x versus normalized horizontal displacement u_x. For a color version of this figure, see www.iste.co.uk/grange/soil.zip*

4.7. Results of the numerical simulations

Several results of the numerical simulations on the behavior of the reinforced concrete viaduct subjected to seismic action are shown in the following, with a comparison of the predictions of both macroelements and a "reference" solution, obtained by constraining displacements and rotations at the base of the piers (where infinitely rigid soil is assumed), in which the effects of SSI are completely ignored.

4.7.1. *Forces and displacements in the structural elements*

Figures 4.20 and 4.21 show the trends of $H_x : U_x$ and $M_y : \Theta_y$ curves for the three pier foundations. Due to the symmetry of the problem, the responses of P1 and P3 piers are identical. The figures show that the predictions of the two macroelements are very similar, although the hypoplastic model provides slightly smaller maximum horizontal displacements. It is interesting to observe that, at greater amplitude cycles, the resistance (to sliding and rotation) of the foundations is fully mobilized, giving rise to irreversible displacements and rotations, associated with relevant hysteresis cycles with related energy dissipation.

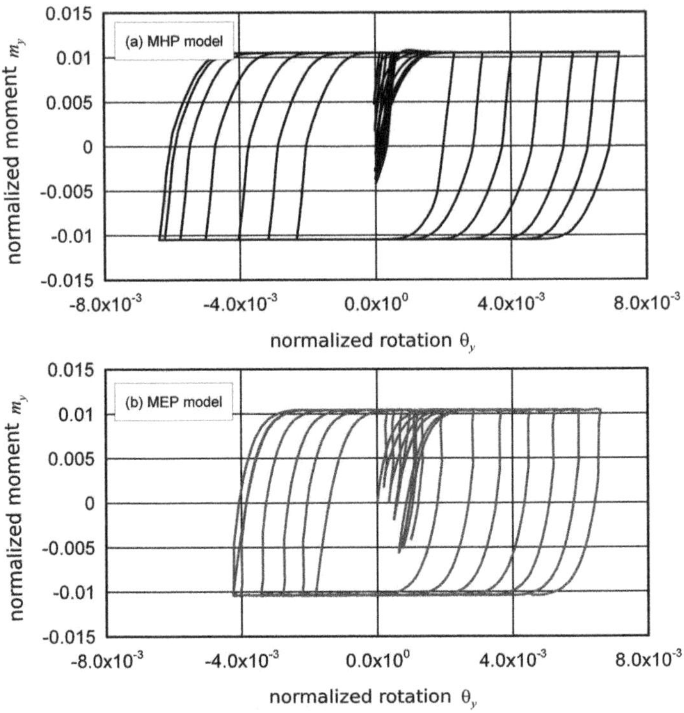

Figure 4.18. *Calibration test no. 3 – phase b): normalized moment m_y versus normalized rotation θ_y. For a color version of this figure, see www.iste.co.uk/grange/soil.zip*

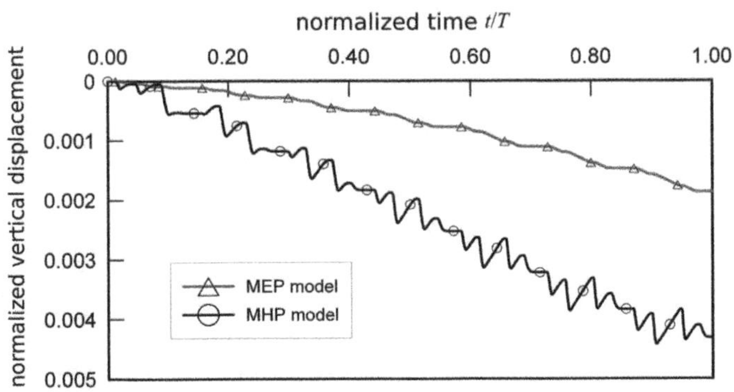

Figure 4.19. *Calibration test no. 3 – phase b): normalized vertical displacement v versus time factor t/T. For a color version of this figure, see www.iste.co.uk/grange/soil.zip*

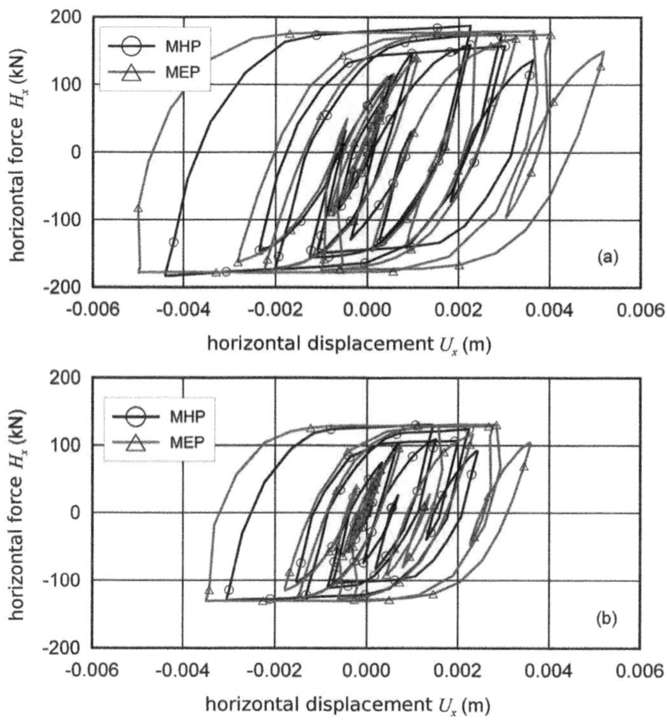

Figure 4.20. *Horizontal force versus horizontal displacement at the foundations: a) P1 and P3 piers and b) P2 pier. For a color version of this figure, see www.iste.co.uk/grange/soil.zip*

The time evolution of the shear force and bending moment at the head of the pier foundations is shown in Figures 4.22 and 4.23 respectively. For the two macroelements, the minimum and maximum values of the shear force H_x and the bending moment M_y have an upper limit represented by the ultimate conditions for sliding and rotation of the foundation; this is not the case for an infinitely rigid soil. While the predictions obtained using the two macroelements are very similar, the reactions calculated in the hypothesis of an infinitely rigid soil are significantly higher, with increments in peak values of about 100%, both in terms of shear forces and bending moment. This result is in agreement with what was observed by, for example, [WOT 10] and [ALG 10] and depends both on the increase in the fundamental period of the structure associated with the deformability of the soil–foundation system and on the energy dissipation due to the hysteretic behavior of the two macroelements.

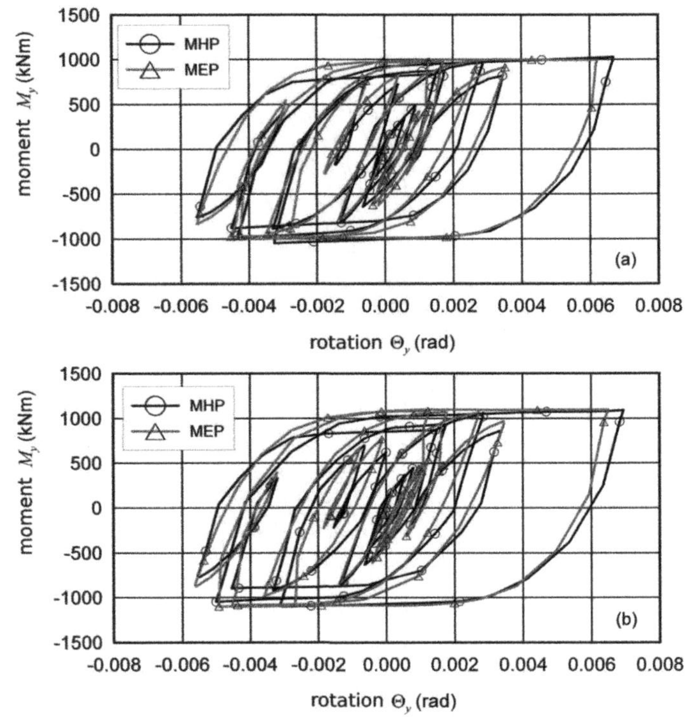

Figure 4.21. *Bending moment versus rotation at the foundations: a) P1 and P3 piers and b) P2 pier. For a color version of this figure, see www.iste.co.uk/grange/soil.zip*

4.7.2. Displacements of the abutment and the foundations

The horizontal displacements at the head of the two piers are shown in Figure 4.24. As seen previously, the solutions obtained with the two macroelements are almost identical. Conversely, the displacements computed for the case of infinitely rigid soil are characterized by a significantly different time evolution.

Contrary to the forces, the peak values of the displacements at the center of the deck – computed considering SSI – are only slightly higher than those obtained in the case of rigid soil. This is due to the fact that, in the latter case, the forces induced by the seismic action applied at the base of the piers are high enough to cause yielding of the pier itself, as shown by the bending moment versus curvature diagram (Figure 4.25). In the analyses conducted considering SSI, the pier remains in the elastic domain for the entire duration of the seismic event.

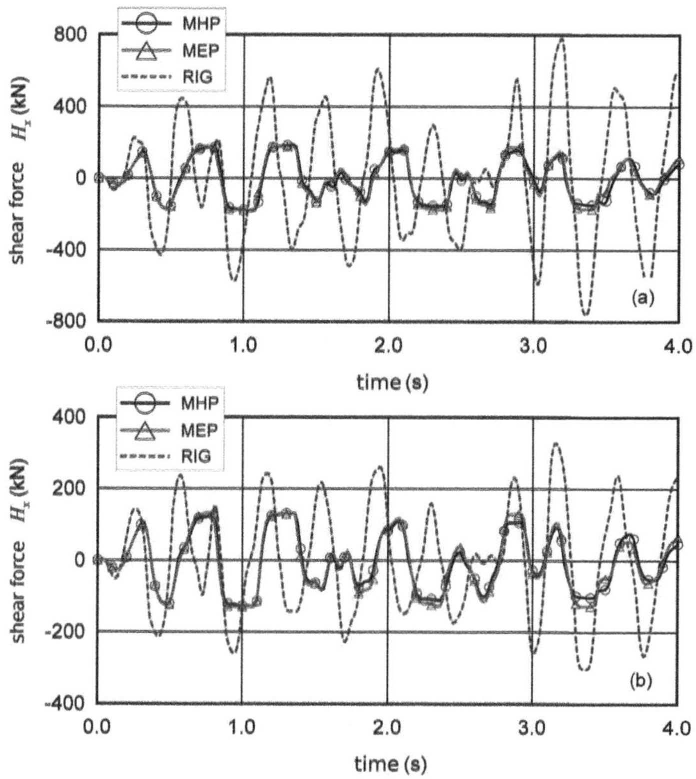

Figure 4.22. *Time evolution of the shear force at the head of the foundations: a) P1 and P3 piers and b) P2 pier. For a color version of this figure, see www.iste.co.uk/grange/soil.zip*

The settlements accumulated by the three foundations during the seismic event are shown in Figure 4.26. The responses provided by the two macroelements show some differences. While the simulation with the elastoplastic macroelement provides very small vertical displacement values (with a maximum of 1 mm at the end of the seismic event) and monotonous increase in time, the settlements accumulated in the simulation with the hypoplastic macroelement are significantly higher (with a maximum of about 6 mm for the P2 pier), although they remain somewhat limited in absolute terms. This difference is likely due to the kinematic feature of the surface defined by the plastic potential function in the model by [GRA 09], whereas in the hypoplastic model, the irreversible deformations are determined from the plastic potential function by [NOV 91], characterized by an isotropic evolution.

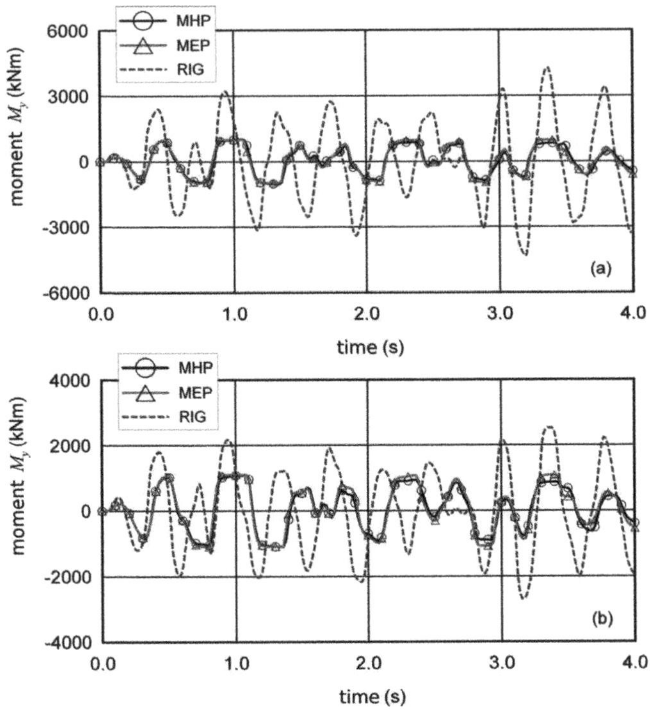

Figure 4.23. *Time evolution of the bending moment at the head of the foundations: a) P1 and P3 piers and b) P2 pier. For a color version of this figure, see www.iste.co.uk/grange/soil.zip*

4.8. Concluding remarks

Two macroelements for shallow foundations subjected to cyclic/dynamic loads, recently developed in the context of the theories of hardening plasticity and hypoplasticity, have been used to evaluate the effects of soil–structure interaction on a pre-compressed reinforced concrete viaduct in seismic conditions.

The simulations carried out using the two macroelements provided very similar results, both in terms of horizontal forces/moments and horizontal displacements/rotations, despite the different mathematical formulations of the implemented constitutive equations. This is probably due to the presence, in both formulations, of an appropriate internal vectorial variable (the internal force τ which in the elastoplastic model assumes the role of "back stress" and the internal displacement δ in the hypoplastic model) which keeps the memory of the previous loading history to allow a description of the hysteretic behavior of the soil–foundation system.

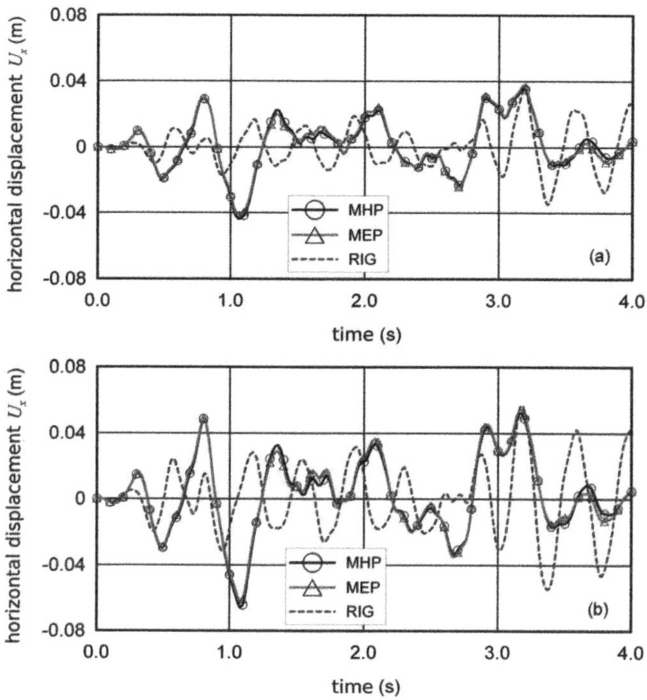

Figure 4.24. *Time evolution of the horizontal displacement of the pier heads: a) P1 and P3 piers and b) P2 pier. For a color version of this figure, see www.iste.co.uk/grange/soil.zip*

The only significant difference is on the magnitude of the settlements accumulated by the foundations under the effect of cyclic loading, which is significantly greater for the case of the hypoplastic macroelement. This is probably due to the different functions of the two macroelements that define the direction of the irreversible component of the velocity ($\dot{\boldsymbol{u}}^p$ or $\boldsymbol{m}$); further studies are, however, currently underway to assess this hypothesis.

The comparison between the macroelement results and those obtained considering an infinitely rigid soil allows the quantitative evaluation of the influence of SSI. For the considered case study, the effect of SSI is beneficial in terms of shear forces and bending moments, which decrease about 50% compared to the rigid soil case. Furthermore, the increase in the deformability of the system does not have a significant impact on the deck displacements. The reason is that the higher forces obtained in the case of rigid soil cause yielding at the base of the piers, resulting in a significant reduction of the structural stiffness.

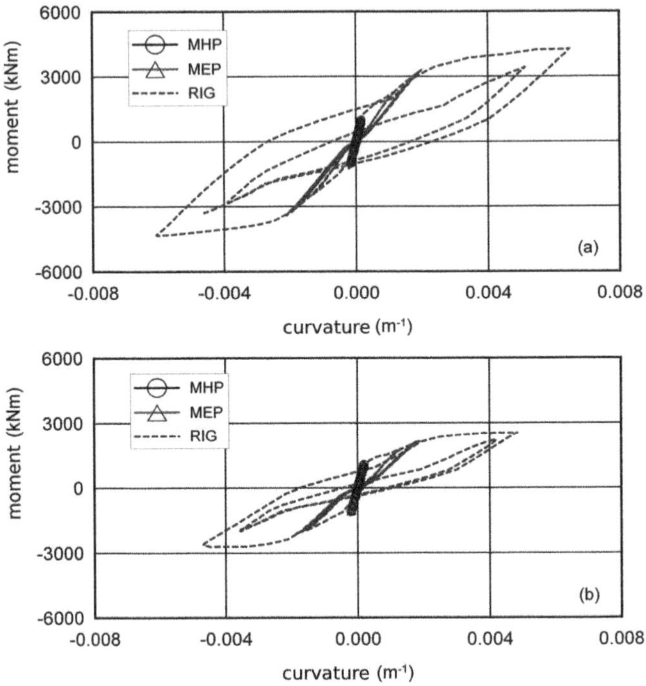

Figure 4.25. *Bending moment versus curvature: a) P1 and P3 piers and b) P2 pier. For a color version of this figure, see www.iste.co.uk/grange/soil.zip*

As observed by [MYL 00], considering soil deformability in the soil–foundation system can lead either to a reduction or to an increase in the computed forces on the structure, depending on the characteristics of the response spectrum of the considered seismic input. Ignoring the effect of SSI in the design stage can therefore lead to inaccurate assessment of the system behavior in a non-conservative side.

The chosen case study illustrates that the two macroelements represent a significant step forward in the development of simple, robust and accurate calculation tools for the quantitative evaluation of the effects of soil–structure interaction in seismic conditions, not only for research purposes but also for structural design. The macroelement concept combines the computational advantages typical of conventional approaches to the ability of reproducing the behavior of soil–foundation systems on complex stress paths, comparable only to that obtained using full 3D finite element models, whose computational cost is prohibitive in everyday practice.

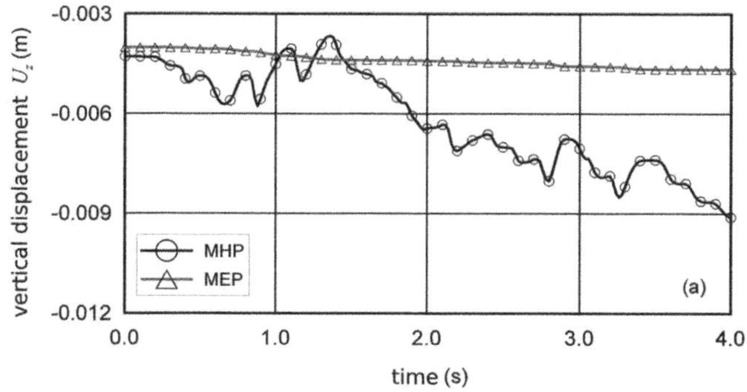

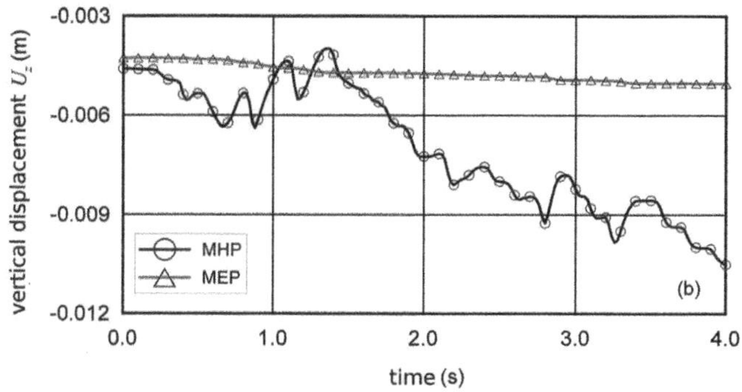

Figure 4.26. *Time evolution of the vertical displacements of the foundations: a) P1 and P3 piers and b) P2 pier. For a color version of this figure, see www.iste.co.uk/grange/soil.zip*

4.9. References

[ALG 10] ALGIE T.B., PENDER M.J., ORENSE R.P., WOTHERSPOON L.M., "Dynamic field testing of shallow foundations subject to rocking", *2010 New Zealand Society for Earthquake Engineering Conference*, Wellington, New Zealand, 26–28 March, 2010.

[BIE 03] BIELAK J., LOUKAKIS K., HISADA Y., YOSHIMURA C., "Domain reduction method for three-dimensional earthquake modeling in localized regions. Part I: Theory", *Bulletin of the Seismological Society of America*, vol. 93, pp. 817–824, 2003.

[BIE 06] BIENEN B., BYRNE B.W., HOULSBY G.T., CASSIDY M.J., "Investigating six-degree-of-freedom loading of shallow foundations on sand", *Géotechnique*, vol. 56, pp. 367–379, 2006.

[BUT 94] BUTTERFIELD R., GOTTARDI G., "A complete three-dimensional failure envelope for shallow footings on sand", *Géotechnique*, vol. 44, no. 1, pp. 181–184, 1994.

[CAQ 66] CAQUOT A., KÉRISEL J., *Traité de mécanique des sols*, Gautiers-Villars, Paris, 1966.

[CAS 02] CASSIDY M., BYRNE B., HOULSBY G., "Modelling the behaviour of circular footings under combined loading on loose carbonate sand", *Géotechnique*, vol. 52, no. 10, pp. 705–712, 2002.

[CAS 04] CASSIDY M., MARTIN C., HOULSBY G., "Development and application of force resultant models describing jack-up foundation behaviour", *Marine Structures*, vol. 17, pp. 165–193, 2004.

[CHA 07] CHATZIGOGOS C., Comportement sismique des fondations superficielles : vers la prise en compte d'un critère de performance dans la conception, PhD thesis, Ecole Polytechnique, 2007.

[CHA 08] CHATZIGOGOS C.T., PECKER A., SALENCON J., "Macroelement modeling of shallow foundations", *Soil Dynamics and Earthquake Engineering*, vol. 29, pp. 765–781, 2008.

[CHO 19] CHOINIÈRE M., PAULTRE P., LÉGER P., "Influence of soil-structure interaction on seismic demands in shear wall building gravity load frames", *Engineering Structures*, vol. 198, pp. 109–259, 2019.

[CIA 95] CIAMPOLI M., PINTO P.E., "Effects of soil–structure interaction on inelastic seismic response of bridge piers", *ASCE Journal of Structural Engineering*, vol. 121, pp. 806–814, 1995.

[CRÉ 01] CRÉMER C., PECKER A., DAVENNE L., "Cyclic macro–element for soil–structure interaction: Material and geometric non–linearities", *International Journal for Numerical and Analytical Methods in Geomechanics*, vol. 25, pp. 1257–1284, 2001.

[CRÉ 02] CRÉMER C., PECKER A., DAVENNE L., "Modelling of nonlinear dynamic behaviour of a shallow strip foundation with macro-element", *Journal of Earthquake Engineering*, vol. 6, pp. 175–211, 2002.

[DAR 78] DARVE F., Une formulation incrémentale des lois rhéologiques. Application aux sols, PhD thesis, INP Grenoble, 1978.

[DAV 73] DAVIS E., BOOKER J., "The effect of increasing strength with depth on the bearing capacity of clays", *Géotechnique*, vol. 23, no. 4, pp. 551–563, 1973.

[DAV 98] DAVENNE L., BRENET C., Macro-éléments de poutres en béton armé, Report, LMT Cachan, no. 210, 1998.

[ELA 92] ELACHACHI S., Sur l'élaboration d'une méthode simplifiée d'analyse des structures de Génie Civil par macro-éléments adaptés aux constructions composites et endommageables, PhD thesis, Université Paris VI, 1992.

[ELG 08] ELGAMAL A., YAN L., YANG Z., CONTE J.P., "Three–dimensional seismic response of Humboldt Bay bridge-foundation-ground system", *ASCE Journal of Structural Engineering*, vol. 134, pp. 1165–1176, 2008.

[ELN 96] ELNASHAI A.S., MCCLURE D.C., "Effect of modelling assumptions and input motion characteristics on seismic design parameters of RC bridge piers foundations", *Earthquake Engineering and Structural Dynamics*, vol. 25, pp. 435–463, 1996.

[EUR 04] EUROPEAN COMMITTEE FOR STANDARDIZATION, Eurocode 8: Design of structures for earthquake resistance, European standard, EN 1998-1, 2004.

[FAR 15] FARDIS M.N., CARVALHO E.C., FAJFAR P., PECKER A., *Seismic Design of Concrete Buildings to Eurocode 8*, CRC Press, Boca Raton, 2015.

[FIL 04] FILIPPOU F.C., CONSTANTINIDES M., *FedeasLab Getting Started Guide and Simulations Examples*, Department of Civil & Environmental Engineering, University of California, Berkeley, 2004.

[GAZ 98] GAZETAS G., MYLONAKIS G., "Seismic soil–structure interaction: New evidence and emerging issues", in DAKOUKAS P., YEGIAN M.K., HOLTZ R.D. (eds), *Geotechnical Earthquake Engineering and Soil Dynamics III*, ASCE, Reston, 1998.

[GOT 99] GOTTARDI G., HOULSBY G.T., BUTTERFIELD R., "The plastic response of circular footings on sand under general planar loading", *Géotechnique*, vol. 49, pp. 453–470, 1999.

[GRA 08a] GRANGE S., Modélisation simplifiée 3D de l'interaction sol-structure: application au génie parasismique, PhD thesis, INP Grenoble, 2008. Available at: http://tel.archives-ouvertes.fr/tel-00306842/fr.

[GRA 08b] GRANGE S., KOTRONIS P., MAZARS J., "A macro-element for a circular foundation to simulate 3D soil–structure interaction", *International Journal for Numerical and Analytical Methods in Geomechanics*, vol. 32, pp. 1205–1227, 2008.

[GRA 09] GRANGE S., KOTRONIS P., MAZARS J., "A macro-element to simulate 3D soil–structure interaction considering plasticity and uplift", *International Journal of Solids and Structures*, vol. 46, pp. 3651–3663, 2009.

[HOU 02] HOULSBY G.T., CASSIDY M.J., "A plasticity model for the behaviour of footings on sand under combined loading", *Géotechnique*, vol. 52, pp. 117–129, 2002.

[JER 04] JEREMIĆ B., KUNNATH S., XIONG F., "Influence of soil–foundation–structure interaction on seismic response of the I-880 viaduct", *International Journal for Engineering Structures*, vol. 26, no. 3, pp. 391–402, 2004.

[JER 09] JEREMIĆ B., JIE G., PREISIG M., TAFAZZOLI N., "Time domain simulation of soil–foundation–structure interaction in non-uniform soils", *Earthquake Engineering Structural Dynamics*, vol. 38, no. 5, pp. 699–718, 2009.

[JIN 19a] JIN Z., YIN Z.-Y., KOTRONIS P., LI Z., "Advanced numerical modelling of caisson foundations in sand to investigate the failure envelope in the HMV space", *Ocean Engineering*, vol. 190, pp. 106–394, 2019.

[JIN 19b] JIN Z., YIN Z.-Y., KOTRONIS P., LI Z., TAMAGNINI C., "A hypoplastic macroelement model for a caisson foundation in sand under monotonic and cyclic loadings", *Marine Structures*, vol. 66, pp. 16–26, 2019.

[KOL 91] KOLYMBAS D., "An outline of hypoplasticity", *Archive of Applied Mechanics*, vol. 61, pp. 143–151, 1991.

[KOT 05] KOTRONIS P., MAZARS J., "Simplified modelling strategies to simulate the dynamic behaviour of R/C walls.", *Journal of Earthquake Engineering*, vol. 9, no. 2, pp. 285–306, 2005.

[KOT 13] KOTRONIS P., TAMAGNINI C., GRANGE S. (eds), *Soil-Structure Interaction, Alert Doctoral School 2013*, Aussois, France, 3–5 October, 2013. Available at: http://alertgeomaterials.eu/publications.

[LAB 91] LA BORDERIE C., Phénomènes unilatérales dans un matériau endommageable: modélisation et application à l'analyse des structures en béton, PhD thesis, Université Paris VI, 1991.

[LEP 01] LE PAPE Y., SIEFFERT J.-G., "Application of thermodynamics to the global modelling of shallow foundations on frictional material", *International Journal for Numerical and Analytical Methods in Geomechanics*, vol. 25, pp. 1377–1408, 2001.

[LI 14] LI Z., KOTRONIS P., ESCOFFIER S., "Numerical study of the 3D failure envelope of a single pile in sand", *Computers and Geotechnics*, vol. 62, pp. 11–26, 2014.

[LI 16] LI Z., KOTRONIS P., ESCOFFIER S., TAMAGNINI C., "A hypoplastic macroelement for single vertical piles in sand subject to three-dimensional loading conditions", *Acta Geotechnica*, vol. 11, no. 2, pp. 373–390, 2016.

[LI 18] LI Z., KOTRONIS P., ESCOFFIER S., TAMAGNINI C., "A hypoplastic macroelement formulation for single batter piles in sand", *International Journal for Numerical and Analytical Methods in Geomechanics*, vol. 42, no. 12, pp. 1346–1365, 2018.

[MAR 94] MARTIN C., Physical and numerical modelling of offshore foundations under combined loads, Master's thesis, DPhil thesis, University of Oxford, 1994.

[MAR 01] MARTIN C.M., HOULSBY G.T., "Combined loading of spudcan foundations on clay: Numerical modelling", *Géotechnique*, vol. 51, pp. 687–700, 2001.

[MAT 79] MATAR M., SALENÇON J., "Capacité portante des semelles filantes", *Revue Française de Géotechnique*, vol. 9, pp. 51–76, 1979.

[MEN 73] MENEGOTTO M., PINTO P.E., "Method of analysis of cyclically loaded reinforced concrete plane frames including changes in geometry and non-elastic behaviour of elements under combined normal force and bending", *IABSE Symposium on Resistance and Ultimate Deformability of Structures Acted on by Well-Defined Repeated Loads*, Lisbon, Portugal, 1973.

[MON 97] MONTRASIO L., NOVA R., "Settlements of shallow foundations on sand: Geometrical effects", *Géotechnique*, vol. 47, pp. 49–60, 1997.

[MUR 84] MURCHISON J.M., O'NEILL M.W., "Evaluation of P-Y relationships in cohesionless soils", *Analysis and Design of Pile Foundations*, ASCE, pp. 174–191, 1984.

[MYL 00] MYLONAKIS G., GAZETAS G., "Seismic soil–structure interaction: Beneficial or detrimental?", *Journal of Earthquake Engineering*, vol. 4, pp. 227–301, 2000.

[NIE 97] NIEMUNIS A., HERLE I., "Hypoplastic model for cohesionless soils with elastic strain range", *Mechanics of Cohesive-frictional Materials*, vol. 2, pp. 279–299, 1997.

[NIE 02] NIEMUNIS A., Extended hypoplastic models for soils, Habilitation thesis, Bochum University, 2002.

[NOV 91] NOVA R., MONTRASIO L., "Settlements of shallow foundations on sand", *Géotechnique*, vol. 41, pp. 243–256, 1991.

[PEC 97] PECKER A., "Analytical formulae for the seismic bearing capacity of shallow strip foundations", in *Seismic Behaviour of Ground and Geotechnical Structures*, SECO E PINTO P.S. (ed.), Taylor & Francis Group, London, 1997.

[PER 20] PEREZ-HERREROS J., Interaction dynamique sol-structure des fondations sur pieux : étude expérimentale et numérique, PhD thesis, Ecole Centrale de Nantes, 2020.

[PHI 03] PHILIPPONNAT G., HUBERT B., *Fondations et ouvrages en terre*, Eyrolles, Paris, 2003.

[PIN 96] PINTO A.V., VERZELETTI G., PEGON P., MAGONETTE G., NEGRO P., GUEDES J., Pseudo-dynamic testing of large-scale R/C bridges, Research report EUR 16378 EN, Joint Research Centre, Ispra, 1996.

[PRI 03] DI PRISCO C., NOVA R., SIBILIA A., "Shallow footing under cyclic loading: Experimental behaviour and constitutive modelling", in MAUGERI M., NOVA R. (eds), *Geotechnical Analysis of the Seismic Vulnerability of Historical Monuments*, pp. 99–121, Pàtron Editore, Bologna, 2003.

[PRI 06] DI PRISCO C., MASSIMINO M.R., MAUGERI M., NICOLOSI M., NOVA R., "Cyclic numerical analysis of Noto Cathedral: Soil–structure interaction modelling", *Rivista Italiana di Geotecnica*, vol. 48, 2006.

[RAN 04] RANDOLPH M., JAMIOLKOWSKI M., ZDRAVKOVI L., "Load carrying capacity of foundations", *Proceedings of the Skempton Memorial Conference on Advances in Geotechnical Engineering*, vol. 1, pp. 207–240, London, United Kingdom, 2004.

[SAL 09] SALCIARINI D., TAMAGNINI C., "A hypoplastic macroelement model for shallow foundations under monotonic and cyclic loads", *Acta Geotechnica*, vol. 4, no. 3, pp. 163–176, 2009.

[SPY 92] SPYRAKOS C.C., "Seismic behavior of bridge piers including soil–structure interaction", *Computers & Structures*, vol. 43, no. 2, pp. 373–384, 1992.

[STO 93] STOER J., BULIRSCH R., *Introduction to Numerical Analysis*, 2nd edition, Springer Verlag, New York, 1993.

[WOL 96] VON WOLFFERSDORFF P.A., "A hypoplastic relation for granular materials with a predefined limit state surface", *Mechanics of Cohesive-frictional Materials*, vol. 1, pp. 251–271, 1996.

[WOT 10] WOTHERSPOON L.M., PENDER M.J., "Effect of shallow foundation modeling on seismic response of moment frame structures", in ORENSE R.P., CHOUW N., PENDER M.J. (eds), *International Workshop on Soil–Foundation–Structure Interaction*, CRC Press, Boca Raton, 2010.

[YOS 03] YOSHIMURA C., BIELAK J., HISADA Y., "Domain reduction method for three–dimensional earthquake modeling in localized regions. Part II: Verification and examples", *Bulletin of the Seismological Society of America*, vol. 93, pp. 825–840, 2003.

[ZHA 02] ZHANG J., MAKRIS N., "Seismic response analysis of highway overcrossing including soil–structure interaction", *Earthquake Engineering and Structural Dynamics*, vol. 31, pp. 1967–1991, 2002.

5

Urban Seismology: Experimental Approach to Soil–Structure Interaction Towards the Concept of Meta-city

5.1. Introduction

Since the beginning of the 19th Century, a large number of authors have been interested in the phenomenon of coupling between soil and foundations under dynamic loading. They addressed these processes by means of empirical, and later analytical, approaches [HAD 93]: for example, Lamb's works in 1904 [LAM 04], which studied the vibrations of a semi-infinite linear elastic medium; Reissner's works in 1936 [REI 36], which highlighted the phenomenon of dissipation of vibration energy in soil in the form of waves; or those of Hsieh in 1962 [HSI 62] and especially Lysmer in 1965 [LYS 65], which for the first time introduced the idea of replacing the soil–foundation system by a simple oscillator, whose stiffness and damping values are constant with frequency.

With the advent of civil nuclear power at the end of the 1950s, the analysis and understanding of *SSI* took on another dimension and led to a new approach. Unlike the first analyses focusing on vibrating machines, seismic interaction considers harmonic loading that originates in soil. While for light structures, *SSI* appeared at first glance to be negligible, rigid or massive structures have shown important interactions, in particular, during deformations in the horizontal directions of structures, and no longer only in the vertical direction.

In the case of a structure based on rocky soil, in other words considered infinitely rigid, the loading wavelength is very large compared to the size of the foundation,

Chapter written by Philippe GUÉGUEN, Philippe ROUX and Andrea COLOMBI.

such that motion at depths, at the soil–foundation contact level, and at the surface are in phase. The motion at the base will therefore directly refer to the center of mass in the form of forces of inertia, causing the deformation of the structure. This deformation will naturally lead to the appearance of a bending moment and a shear force that represent the reaction of the non-deformable rock. No soil deformation is to be expected at the base of the foundation and free-field motion is not disturbed by the presence of the foundation. It will then be said that the structure is fixed or rigid.

In contrast, we will refer to a flexible-base structure when the latter has its foundations in soil, soft enough to allow a relative shift of the foundation with respect to soil. This situation is comparable to the case where heterogeneities can cause perturbations in the wave field, from the moment their impedance contrast with soil becomes significant. For example, irregularities located at the level of soil-substratum contact can cause interactions between surface waves and volume waves [HIL 84, LEV 85, CHA 90], as well as at the soil surface [WIR 89, CHA 90]. Since the foundation can be considered to be rigid compared to the soil, the kinematic compatibility between soil and foundation deformation will induce a phase shift between incident and diffracted motion, all the more significant as the wavelength of the incident field will be at least of the same order of magnitude as the size of the foundation [KAU 10]. The interaction introduces scattering and resonance phenomena along the length of the foundations [STE 98], which can pollute the incident wave field. This effect is exacerbated by the presence of a strong impedance contrast between soil and foundations [TRI 72, WON 77, MOS 87].

On the contrary, the inertia developed in the structure causes translation and rocking motion of the foundation (and therefore of the structure) relative to the soil. Given that soil is flexible, it will deform under the action of the efforts transmitted by the foundation. As a result, an additional deformation field will propagate into the soil in the form of seismic waves [JEN 70, KAN 71, GUÉ 00, COR 04, KIM 01].

This shows that significant couplings occur between soil and structure, as soon as they are subjected to seismic load. Because of the kinematic or inertial effects, the incident wave field is perturbed and moves away, with these characteristics, from the seismic motion that would be observed over a space free of any constructions. The diffracted wave field at the foundation level, the seismic motion caused by structure vibration, and the interactions between structures, suggest that the seismology of farmlands appears very different from the seismology of cities (or urban seismology).

In this chapter, we will illustrate with observations, the phenomenon of coupling that may exist between soil and structure and its consequences on the dynamic response of the soil–structure system and the nearby seismic ground motion. First, we will show some examples of meaningful interactions. Then, we will describe the interactions observed between structures under dynamic loading. Finally, we will illustrate at the city scale the generalization of site–city effect by introducing the concept of metamaterials applied to urban seismology.

5.1.1. *Observation of soil–structure interaction under weak and strong seismic loading*

5.1.1.1. *Analytic model for a 1-DOF system*

Among the simplified methods, Richart and Whitman [RIC 67] in 1967 extended Hsieh's [HSI 62] and Lysmer's works [LYS 65] to structural vibrational motion in general. Concerning the translation modes in all three directions and the rotation modes around the three main axes, and following the same philosophy as for the macroelements of Chapter 4, they proposed static stiffness and damping formulas. A pivotal period in the evaluation of *SSI* corresponds to the publication, at about the same time, of two articles, [LUC 71, VEL 71], which expressed the solution to the dynamic response of a circular foundation based on an infinite and elastic half-space, subjected to an external harmonic force. They established this solution for a massless foundation, using an expression that connects displacements to efforts, for vertical, horizontal, flexion and torsion harmonic forces: the impedance functions K are the functions that depend on elastic parameters of soil and on the geometric dimensions of the foundation.

Different methods for calculating impedance functions are presented in [GAZ 91]. They are grouped into four categories: analytical or semi-analytical methods, dynamic methods by numerical methods, hybrid methods combining analytical and numerical approaches, and approximated methods that make it possible to simply solve singularities that cannot be rigorously obtained. A wealth of literature is available regardless of the types of soil and foundation being considered e.g. [VEL 71, LUC 71, KAU 74, LUC 74, GAZ 83, APS 87]. Another approach consists of defining analytical cone models that represent the impedance conditions of a foundation on soft soil [WOL 85].

A 1D model is commonly used to represent the response of the soil–structure system [LUC 88], (Figure 5.5). Under seismic loading, the total horizontal movement of the system x_{tot} is given by

$$x_{tot} = x_1 + x_0 + H\phi + x_g \qquad [5.1]$$

where x_g is the soil displacement, x_0 is the horizontal movement of the foundation relative to the soil, $H\phi$ is the horizontal displacement at the top produced by the rocking motion of the foundation around the axis perpendicular to the horizontal direction, with H being the height of the structure and x_1 being the bending motion of the fixed-base structure. In this equation, the vertical displacement of the structure is overlooked, and a uniform rocking motion is assumed over the height of the structure.

By neglecting the coupling between the translation motion and rocking, and assuming that the soil does not have rotation motion, the system response is expressed by [GUÉ 05]:

$$
\begin{aligned}
m_1(\ddot{x}_1 + \ddot{x}_0 + H\ddot{\phi}_y) + c_1\dot{x}_1 + k_1 x_1 &= -m_1\ddot{x}_g \\
m_0\ddot{x}_0 + c_h\dot{x}_0 + k_h x_0 + c_1\dot{x}_1 - k_1 x_1 &= -m_0\ddot{x}_g \\
J_0\ddot{\phi}_y + c_r\dot{\phi}_y + k_r\phi_y - Hc_1\dot{x}_1 - Hk_1 x_1 &= 0
\end{aligned}
\quad [5.2]
$$

with m_1, m_0 and J_0 being the masses of the structure and the foundation, and the moment of rotational inertia of the foundation; c_1 and k_1 being the viscous damping and the rigidity of the fixed-base structure; c_h, k_h, c_r and k_r being the damping and stiffness coefficients of the impedance functions for translation h and rocking r movements.

A conventional approach for identifying the response of a soil–structure system is to determine from equation [5.2] the resonance frequencies and dampings associated with the vibration modes of the system. Stewart and Fenves [STE 98] summarized the input/output pairs of the system (Table 5.1) that ultimately represent the information required for its experimental identification: we need recordings of free-field motion, of foundation motion and those at the top of the construction to enable the components of the motion of the system to be directly obtained, similarly to [GUÉ 05].

System	Input	Output
Flexible base	x_g	$x_g + x_0 + H\phi + x_1$
Pseudo-flexible base	$x_g + x_0$	$x_g + x_0 + H\phi + x_1$
Fixed base	$x_g + x_0 + H\phi$	$x_g + x_0 + H\phi + x_1$

Table 5.1. *Components of the motion of the soil–structure system, measured for experimental evaluation, according to [STE 98]*

The effect of the stiffness contrast between soil and structure is illustrated with a simple analytical model, which solves analytical equation [5.2] by considering the expression of impedance functions for a superficial foundation resting on different types of soil (Figure 5.1). The results illustrate observations made on numerous real buildings [STE 99], which showed that in most cases the natural frequencies of structures based on soft soil (flexible base) are lower than in those on rigid soil (fixed base). In some cases, it is also found that the frequency of the soil–structure system is located below the maximum energy of seismic load, resulting in a decrease in efforts in the structure. Finally, the efforts at the base of the foundation are exacerbated in the case of rigid soil.

In this model, the kinematic interaction is considered as being a second-order effect. It will only have consequences for exceptional structures with long, deeply buried foundations. We also find that the damping of a structure increases when it is on a soft soil. In the case of rigid soil, damping is caused by the loss of energy during cycles due to internal friction and nonlinearity. On the contrary, in the case of soft soil, some of the vibration energy is dissipated in the form of seismic waves emitted in soil at the soil–foundation contact level, being the result of inertial interaction. This additional damping has a positive effect on the safety of the structure, by reducing the maximal amplitude of deformations and the duration of oscillations.

5.1.1.2. *Frequency analysis of motion components of the soil–structure system*

Based on equation [5.1], Stewart and Fenves [STE 98] have interpreted the soil–structure system response, considering several configurations. If the structure and soil are flexible, the fundamental frequency of vibrations observed at the top of the building corresponds to the response of the system f_{sys}. This frequency is different from the frequency of the fixed-base structure f_1, which corresponds to a flexible structure on rigid soil. On the rigid soil, the fixed-base structure and the system vibrate at the same frequency, namely $f_{sys} = f_1$. In this case, the soil–foundation anchoring condition is perfect, and soil and foundations move in the same way. Luco *et al.* [LUC 88] then proposed approximate equations linking the components of the soil–structure system, i.e.

$$\frac{1}{f_{sys}^2} = \frac{1}{f_1^2} + \frac{1}{f_\phi^2} + \frac{1}{f_0^2} \qquad [5.3]$$

with f_0 and f_ϕ being the frequencies of the horizontal translation motion and foundation rocking in relation to soil. Equation [5.3] indicates that the system parameters depend on the fixed parameters and on the impedance of the foundation in translation and rocking. This equation also implies that if the building is very rigid ($f_1 = \infty$), the system frequency f_{sys} is equivalent to the frequency of the rigid body. If soil is very rigid, f_{sys} is thus equal to the fixed base frequency f_1. An exceptional situation also exists for a partial flexibility condition of the foundation, representing rocking only, i.e. assuming very rigid foundations in the horizontal direction. This so-called apparent frequency f_{app} (or pseudo-flexible system) is given by

$$\frac{1}{f_{app}^2} = \frac{1}{f_1^2} + \frac{1}{f_\phi^2} \qquad [5.4]$$

The parameters of the flexible-base structure can thus be approached by those of the pseudo-flexible model. According to this model, and in the case of a rigid building on soft soil, f_1 is always larger than f_{sys}. This implies that the ratio of structure rigidity to soil is the key factor controlling the lengthening of the frequency [BIE 75, AVI 96].

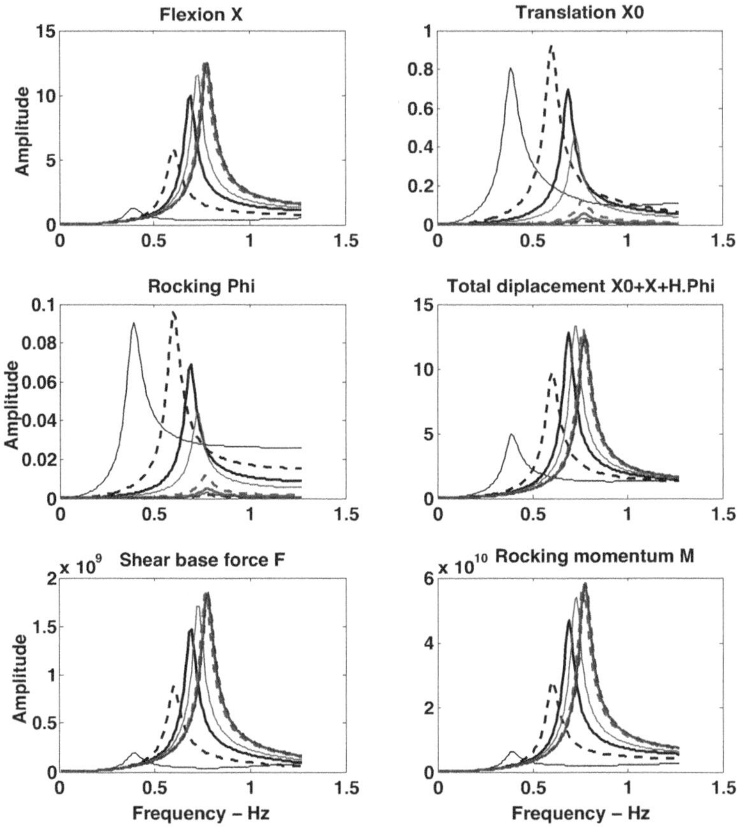

Figure 5.1. *Example of the effect of soil–structure interaction on the response of a structure with a vibration frequency of 0.75Hz, according to the shear wave velocity of the soil. The motions and the efforts correspond to those of a model with one degree of freedom [GUÉ 05], including a structure, foundation and soil described by equation [5.2]. Efforts are obtained directly from the classic relation connecting them to foundation displacements via impedance functions. Black continuous line: Vs=100 m/s; black dashes: 150 m/s; black bold: 200 m/s; thin continuous red: 400 m/s; red dashes: 600 m/s; red continuous bold: 800 m/s; blue: 1000 m/s. For a color version of this figure, see www.iste.co.uk/grange/soil.zip*

By analyzing the variations in frequencies (in particular, f_1 and f_{sys}), the contribution of *SSI* can be observed on the system response, as well as nonlinear phenomena that appear according to loading levels. For example, Trifunac *et al.* [TRI 10] observed, on a reinforced concrete building, a stable frequency f_1 for moderate loading levels, while f_{sys} varied from about 10-15%, indicating a strong sensitivity of the interaction with the loading level. This observation is also reported by Michel *et al.* [MIC 10b] with the Grenoble City Hall, France, where the presence of *SSI* is identified by analyzing the difference between f_1 obtained under weak earthquakes and f_{sys} obtained by processing recordings of ambient vibrations. However, by compiling data from different earthquakes of varying significance, classified according to the maximal structural drift produced between the top and base of the building, and unlike Trifunac *et al.* [TRI 10], Michel *et al.* [MIC 10b] observed for deformations limited to 10^{-5} variations in frequency f_1 of the order of 2%, indicating a structural nonlinear effect under weak loading. This variation is due to the opening and closing during loading of pre-existing cracks, which locally alter stiffness, as other authors also reported [CLI 06, MIC 10a, GUÉ 16b]. Still others have also suggested elastic nonlinear (ENL) behaviors to explain variations in elastic properties under low deformation levels. Characterized at the laboratory scale with rock samples [GUY 99, JOH 05], this ENL behavior is the signature of a heterogeneity rate existing in the system under test (in this case, different rock samples). In a California building, this ENL behavior has recently been demonstrated through the analysis of variations in elastic properties (frequency and damping) over time [GUÉ 16b]. This observation reflects physical processes that are currently poorly constrained, occurring in soil, in the structure or at the interface, which must be better defined to interpret these variations, in terms of structural health.

Todorovska *et al.* [TOD 08] analyzed the ratio of stiffnesses f_1 and f_{sys} under earthquakes of stronger magnitude (e.g. San Fernando or Northridge) observed in the Van Nuys building in Los Angeles. Their analysis showed that during the San Fernando earthquake, f_1 decreased by about 40% and by about 22% during the Northridge earthquake; at the same time, they also found that, although the first frequency of the system f_{sys} was always smaller than f_1, the difference between these two characteristic frequencies varied, depending on the magnitude of the seismic loading. Nonetheless, the model under consideration for extracting the components of the *SSI* was linear, reflecting a particular behavior at the soil–structure interface.

In this study, they applied an alternative approach to the analysis of the system response in the time domain, based on the analysis of the propagation of deformation [ŞAF 98, ŞAF 99, SNI 06]. By calling it Seismic Interferometry by Deconvolution (SIbyD), Snieder and Safak [SNI 06] considered the structure as a one dimensional layer, in which a pulse wave obtained by deconvolution is propagated. In other words, it is tantamount to performing the deconvolution of the top signal by the bottom signal. This results in the structural impulse response, i.e. its frequency transfer function. Mehta *et al.* [MEH 07], Nakata *et al.* [NAK 12], Chandra *et al.* [CHA 15] and Guéguen [GUÉ 16a] applied the SIbyD method successfully with columns of soil instrumented by vertical accelerometer networks to obtain elastic properties and soil response (particularly the shear wave velocity Vs).

This method is particularly accurate for detecting small fluctuations in the building properties. Moreover, Guéguen *et al.* [GUÉ 17b] identified using this method, the effect of changing the anchoring conditions of a structure, by analyzing the ratio of stiffnesses between the structure and the soil–structure system (Figure 5.2). The change was the consequence of a period of extreme cold, and although the depth of penetration of temperature diffusion remains superficial, they showed that the soil–structure system became stiffer when the top layer of the soil was frozen. In this case, we can observe a transition from the structure with flexible behavior to a pseudo-flexible behavior, which can have moderate effects during earthquakes, but affects the assumption of invariance over time, often applied to *SSI*, with important implications for the development and adjustment of realistic models for damage detection and for the physical interpretation of frequency variations observed in actual buildings.

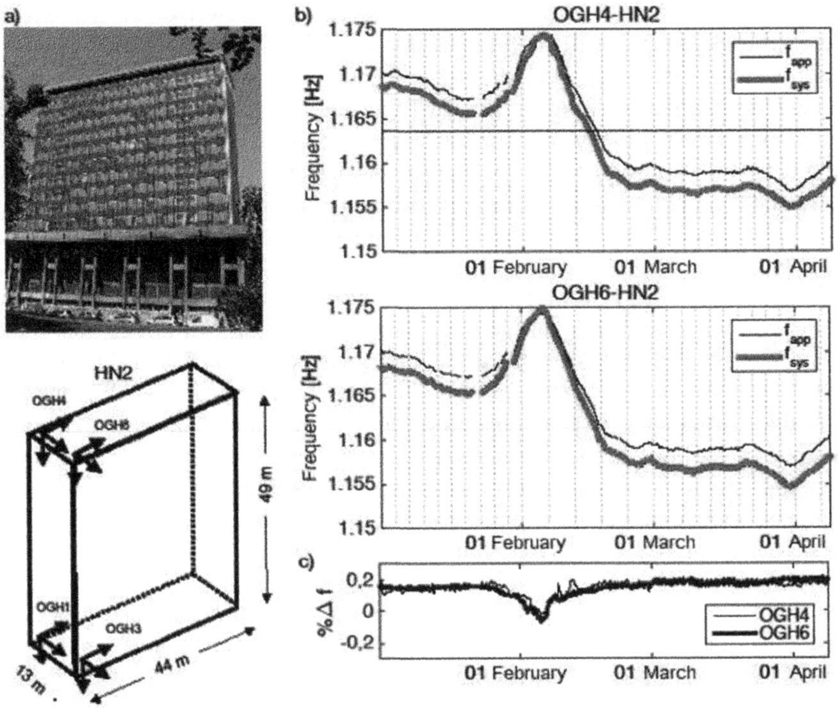

Figure 5.2. *a) Presentation of the Grenoble Town Hall (France) and the permanent instrumentation used in this study. b) Seasonal variation in the frequency of the system f_{sys} (in red) and in the pseudo-flexible frequency f_{app} for 100 days in early 2012 in the longitudinal direction (HN2 component) observed at the OGH4 and OGH6 stations located at the top of Grenoble City Hall (France). c) Seasonal variation in frequency variation Δ_f (in %) calculated between f_{sys} and f_{app} ($\Delta_f = (f_{app} - f_{sys})/f_{app}$), according to [GUÉ 17b]. For a color version of this figure, see www.iste.co.uk/grange/soil.zip*

5.1.1.3. *Cross-interaction between soil and structure*

The experimental analysis of the structure and soil response generally does not make it possible to easily capture the deformation induced by shear located under the surface foundations. However, a soil-induced nonlinear response can be expected when buildings are subjected to heavy loading and when the constraint under cyclic loading increases in the upper layers under the foundations. To analyze this phenomenon, one solution consists of reproducing soil–structure systems in centrifuge. These dynamic experiments have proved to be effective in analyzing the propagation of waves in soil [SEM 98] and the structure–soil and structure–soil–structure coupling under seismic excitation [GHO 06, MAS 13, CHA 03, TRO 13, TRO 14]. Furthermore, Bray and Dashti [BRA 14] have recently used centrifuge models to analyze the vibration-induced effect of the structure on liquefaction localized just below the foundations. At the same time, again using the SIbyD method, Chandra *et al.* [CHA 16] evaluated the nonlinear response of the soil by analyzing variation in shear wave velocity and associated deformation, using a network of accelerometric sensors buried in the centrifugated soil medium.

Petrovic and Parolai [PET 16] applied the SIbyD method to recordings acquired by sensors installed in nearby buildings and boreholes to study the interactions between soil and structure, by analyzing wave propagation through the building-soil layers. In this study, the analysis shows complicated patterns of wave propagation in the soil, resulting from coupling between soil and building. By deconvolution, they separated the various components of the total wave field, in particular by separating the incident wave from the backscattered wave in soil. They thus estimated that the energy radiated by the structure at different depths, immediately under the foundation, could be in the order of 10% of the energy of the incident wave field. This observation confirms the results of Guéguen *et al.* [GUÉ 00] and Guéguen and Bard [GUÉ 05] who, for the first time, interpreted the energy ratios between the vibration of the structure and the wave backscattered in soil. They noted that:

– the time damping of the induced soil motion is proportional to the damping of the soil–structure system, at its fundamental frequency. As a result, part of the vibration energy of the structure is backscattered into soil in the form of a seismic wave, which is consistent with what is known about the effects of *SSI*;

– the spatial decay of the wave field is successively proportional to the geometric damping of volume waves (namely $1/r$ in the immediate vicinity) and surface waves $(1/\sqrt{r})$ from a certain distance, equal to about five times the dimensions of the foundation;

– the vibration of the structure is detectable up to about 10 times the size of the foundation, at which distance the ground motion still represents 5% of the motion at the base of the building. At a distance of twice the size of the foundation, this proportion reaches 25%.

The energy released into soil in the form of seismic waves can therefore be very significant. The waves thus emitted can alter soil motion near the structure, interacting with these neighbors and questioning the representativeness of free-field motion for applications in urban environments.

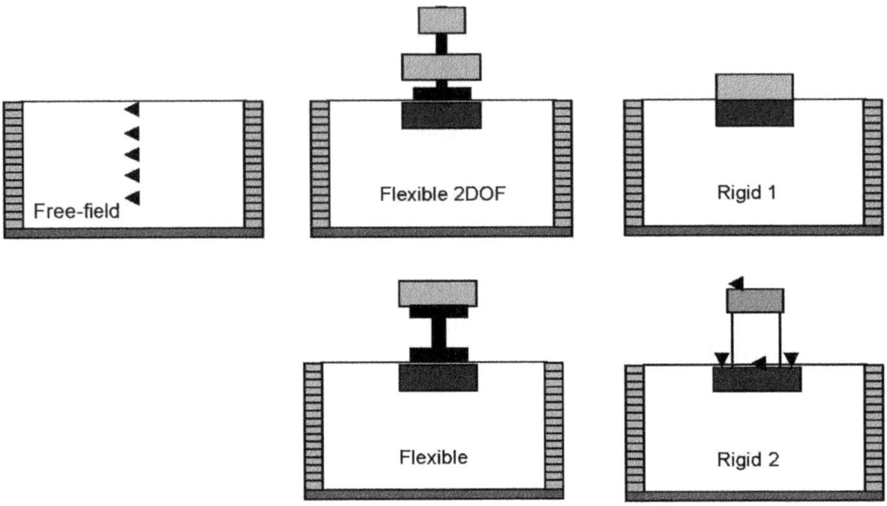

Figure 5.3. *Models of structures tested in centrifuges. The black arrow indicates the position of the sensors used in the medium to calculate the response of soil column by SIbyD, according to the different configurations (either in free-field or with the different structures being tested), and in the structure to calculate the components of the motion of the soil–structure system, according to [CHA 19]*

Resuming the centrifuge tests, Chandra and Guéguen [CHA 19] tested, also by SIbyD, cross-interactions between soil and structure, for stand-alone structures of different configurations (rigid, flexible, with two degrees of freedom, etc. – Figure 5.3), and depending on the loading level. Their main conclusion is as follows:

– It seems that the errors observed between the frequencies obtained by the analytical equations of Luco *et al.* [LUC 88] (equations [5.3] and [5.4]) and the experimental approach are all the more important because the structure is rigid in relation to soil (Figure 5.4b). In this case, the interaction increases, notably due to the increasing rocking mode: the cyclical nature of the load, nonlinearity in soil and at the soil–structure interface induce a bias in the identification of the *SSI*. The same observation is made when the fixed-base frequency is higher, which increases the impedance contrast between soil and the structure, and thus the *SSI* effect. The presence of buildings changes the response of the site, as well as its nonlinear response (Figure 5.4a). This change is illustrated by the reduction of the nonlinear response of the soil when buildings are present. This reduction may originate due to rocking motion, which cyclically increases the containment pressure and reduces soil deformation. This phenomenon is more pronounced for rigid buildings.

– The rocking motion dominates in the total structure motion and is the most sensitive component to nonlinearity.

– It is observed that soil nonlinearity is one of the parameters affecting the nonlinear response of the system, with a more significant impact on the rocking motion.

– Nonlinear *SSI* is more pronounced for rigid buildings than for flexible buildings.

All these observations challenge the validity of soil models derived from free-field conditions for urbanized areas, as well as modifications in site conditions after construction. The effects can be non negligible: the presence of buildings can reduce by 2 the frequency offset in the soil transfer function between weak and strong motion. They can also call into question the reliability of the nonlinear soil model, which is defined without taking into account the presence of buildings. In order to improve the prediction of the building response for design purposes, the site response must incorporate the effects due to the presence of structures into the nonlinear soil response and soil–structure interaction. Nonetheless, soil dynamic loading produced by the motion of the building itself due to rocking must also be integrated, which can increase cyclic loading in the top layer. In the presence of saturated soil, the response could be different, since rocking may accentuate liquefaction [BRA 14]. This observation also suggests that in addressing the problems of nonlinear *SSI*, soil and structure should be considered as a single system, presenting multiple and cross-interactions.

5.1.1.4. *Interactions between structures*

The heterogeneity of urbanizations also very often cause several types of buildings, of different sizes, foundation systems and structural elements, to exist side by side. The distribution of the building then generates large lateral variations and significant contrasts between neighboring structures. Imposing masses are thus built next to lighter structures. In dynamics, it is legitimate to consider whether the presence of large vibrating masses near light masses does not create behavioral changes or even dynamic interactions between buildings.

Kitada *et al.* [KIT 99] studied the interaction effects between buildings at an experimental site. The study consisted of causing a reference structure to vibrate using an exciter placed at the top. This experiment was reproduced in centrifuge by Chazelas *et al.* [CHA 03], as well as by using computers in the context of this study (Figure 5.5). In these three examples, the analysis consisted of assessing the response of a structure, with or without the presence of a nearby building, excited by a load applied at its top (Table 5.2).

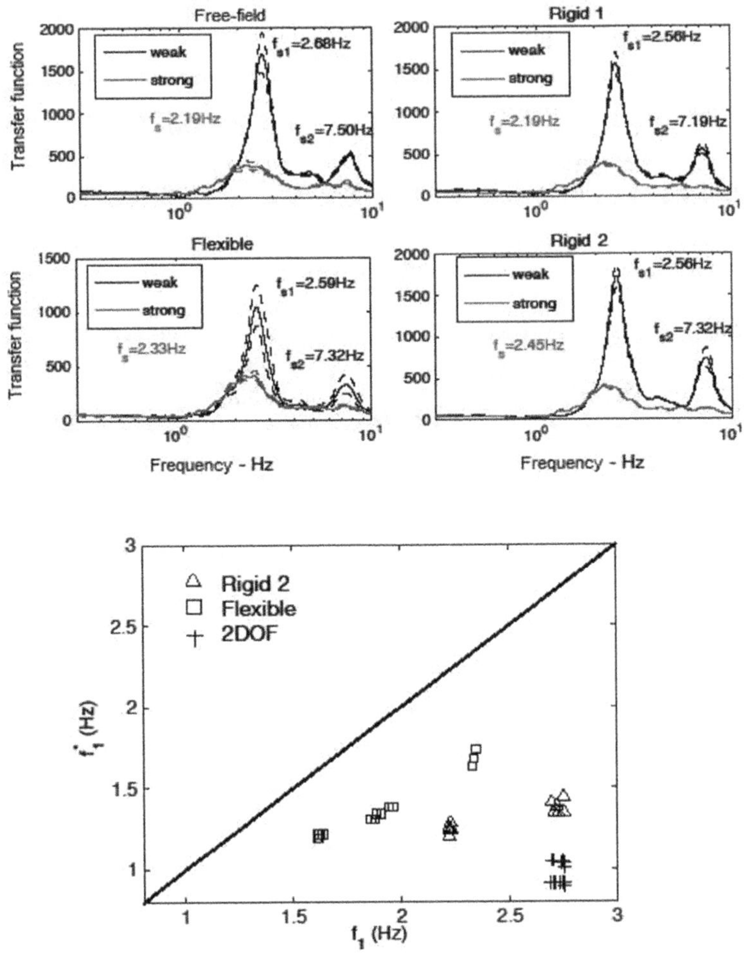

Figure 5.4. *Results of centrifuge trials, according to [CHA 19]. a) Soil column response (SIbyD method), under light (in black) and heavy (in red) loading for different configurations, with or without buildings. The frequency values correspond to the resonance frequency of the soil. b) Comparison of the fixed-base frequency extracted by deconvolution (SIbyD method, f_1^*) or by the equations of the model equation [5.3] (f_1). The continuous line corresponds to the 1:1 relation. For a color version of this figure, see www.iste.co.uk/grange/soil.zip*

In all these experimental examples, numerical studies or in centrifuge, the same observations are reproduced. First, it can be noted that there is a strong coupling between the active structure and the passive structure, especially when the resonance between the soil and the active building exists and when the buildings have similar resonance frequencies (Figure 5.5). The essential effect is to cause the passive building to oscillate, in the order of 10% of the one calculated at the top of the active building, even for significant distances (in the order of the height of the building). It can also be noted that this phenomenon exists for the higher modes of the structure. Since the frequency of the active building and that of the passive building are close to one another, the shape of the beatings is characteristic of the coupling phenomena of resonant systems with almost identical frequencies. In frequency (Figure 5.6), this is reflected by a decoupling of the vibration frequency of the structure, as shown by Kitada *et al.* [KIT 99].

Building	B (m)	H (m)	D (m)	f_1	ζ_1 (%)
B1E	10	20	10	*1.64*	*5.2*
B1S			0	*1.67*	*5.8*
B2E	20	40	10	*0.88*	*5.4*
B2S			0	*0.90*	*5.8*
B3E	40	40	10	*1.25*	*6.2*
B3S			0	*1.32*	*6.5*
B4E	40	50	10	*0.93*	*5.8*
B4S			0	*0.98*	*6.1*
B5E	40	60	10	*0.70*	*12.0*
B5S			0	*0.74*	*12.7*

Table 5.2. *Characteristics of modeled structures. B, H and D respectively represent the width, height and recess of the buildings. f_1 and ζ_1 are respectively the frequency and damping of fixed-base structures, inferred from modeling*

Profile	Layer	$\beta_s(m/s)$	$\rho(t/m^3)$	$E(10^8 Pa)$	ν	$z(m)$
I	1	100	1.50	0.38	0.25	0
III	1	200	1.50	1.50	0.25	0
	2	250	1.80	2.81	0.25	25
	3	300	2.00	4.50	0.25	45
	4	330	1.95	5.31	0.25	85
	5	460	2.10	11.0	0.25	130
	6	600	2.00	18.0	0.25	195

Table 5.3. *Characteristics of the 1D stratified half-space used in the model. β, ρ, E and ν respectively correspond to the velocity of shear waves, to the density, the elasticity modulus (or the Young modulus) and to the Poisson ratio of the different layers. z represents the position of the roof of the layer*

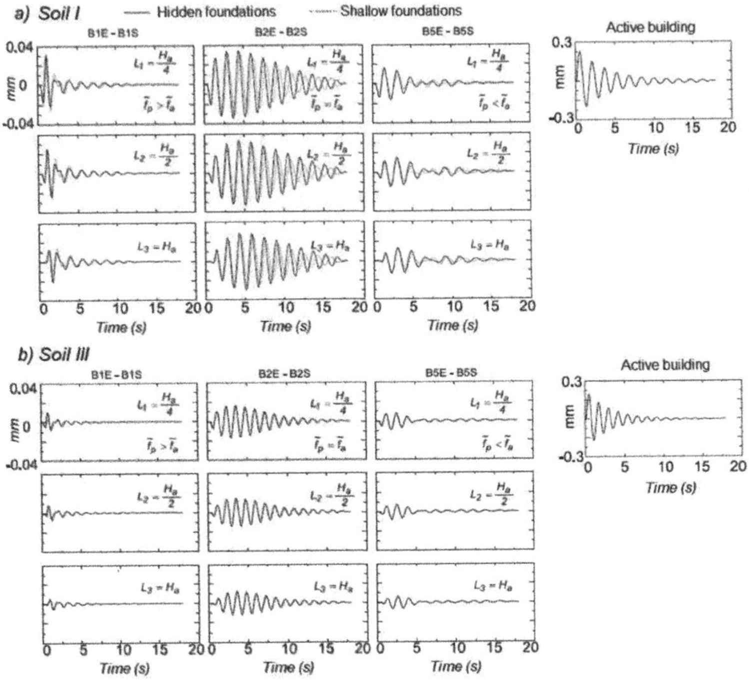

Figure 5.5. *Examples of displacements $u(t)$ calculated at the top of buildings B1E-B1S, B2E-B2S and B5E-B5S for the cases where they have foundations on (a) type-I soil and (b) type-III soil (Table 5.2). Three distances were tested, which correspond to $L_1 = H_a/4$, $L_2 = H_a/2$ and $L_3 = H_a$. The classes of buildings presented correspond to the situations where $\tilde{f}_p > \tilde{f}_a$ (B1E-B1S), $\tilde{f}_p \approx \tilde{f}_a$ (B2E-B2S) and $\tilde{f}_p < \tilde{f}_a$ (B5E-B5S). For example, on the right, we can see the displacement $u(t)$ of the active building, for the case of the passive building B2E based on soil I (on top) and III (at the bottom)*

In 2014, an earthquake of magnitude 4.9 occurred 100 km south of Grenoble [COU 14]. An accelerometer station in the French permanent accelerometer network [PÉQ 08], located in the vicinity of three large buildings in Grenoble (Figure 5.7), constituting a cluster, recorded a horizontal soil acceleration of $2.5cm/s^2$. The Y-shaped sedimentary basin of the Grenoble Valley, composed of thick glaciolacustrine deposits (up to 1 km) systematically amplifies seismic motion [LEB 01, GUÉ 07] between 0.3 and 5.0 Hz, with the result that the inhabitants located in the three towers felt the vibrations more strongly. This was particularly the case during the 2014 earthquake. An investigation into what inhabitants had experienced was immediately conducted and revealed that within the cluster, the inhabitants in the central tower had hardly felt the vibrations of the earthquake, compared to those of the two towers at the extremity of the cluster. Guéguen and Colombi [GUÉ 17a] thus showed that, while the three towers were identical and with foundations in the same soil, not only was the modal response of the middle structure different from that of the other two, but also the seismic response of the structure alone or included in the cluster differed considerably. This difference was particularly sensitive to the position of the seismic source, in relation to the orientation of the cluster (Figure 5.7).

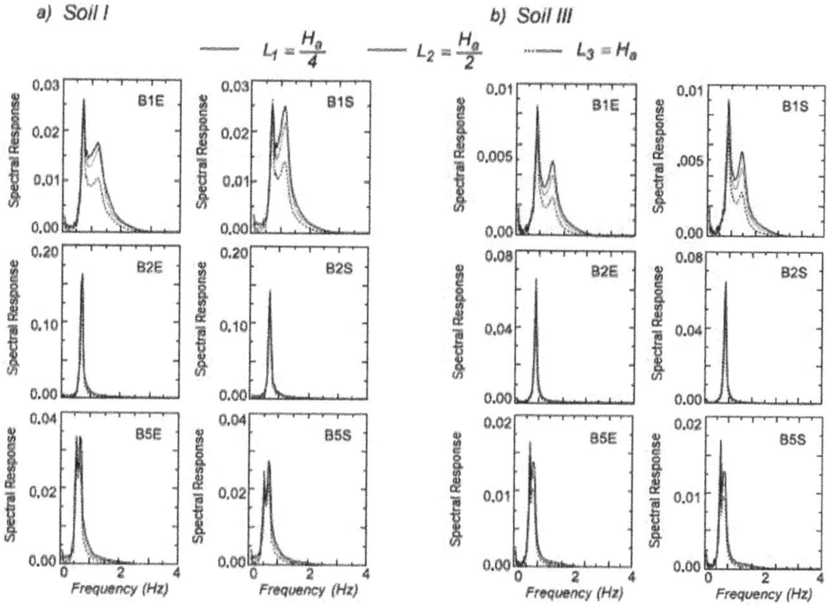

Figure 5.6. *Spectral responses of soil–structure systems presented in Figure 5.5*

This observation ultimately validates previous centrifugal and modeling tests, and confirms the observations made by Kitada *et al.* [KIT 99]. It also questions the meaning of the response of a structure generally defined without taking into account its nearby environment and the ability of a structure to interact with these neighbors, under seismic loading. What has been observed in the elastic domain can have an exacerbated impact when stronger motions occur, influencing the relation between the parameter characterizing the engineering seismic demand parameter EDP (e.g. inter-stage deformation) and the intensity measure IM of seismic motion (e.g. the maximal value of acceleration). This relation between EDP and IM is essential in seismic design based on performance [PRI 00]. The difference in energy between the case of a stand-alone structure, or located within a cluster, can reach 20%, depending on the azimuth of seismic loading, amplifying or mitigating the building motion. This example only concerns three buildings; however, it is reasonable to imagine that global coupling at the urban scale can lead to an even more complex response that can be characterized analogously to a regionally-sized geophysical material. Composed of a group of resonators, it can reveal a new type of physics, which in our case could control wave propagation within designated frequency ranges. When observing the very structured organization of certain exposed cities (Figure 5.8), the question of how these resonators composed of cities would react, in the event of an earthquake, ought to be raised.

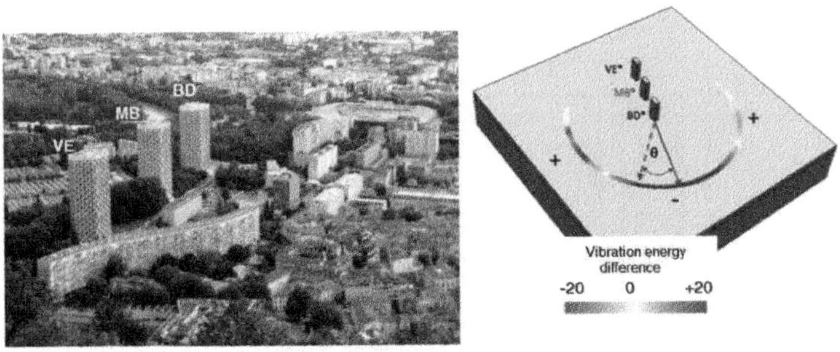

Figure 5.7. *Variation in the difference of vibration energy of the central tower (MB*) between the case where it is isolated and the case where it is inside the cluster of the three towers, for different azimuth values of the source θ with respect to the alignment of the cluster. + and - indicate the positive or negative energy differential. For a color version of this figure, see www.iste.co.uk/grange/soil.zip*

5.1.2. *Contamination of urban seismic motion by the vibration of buildings*

In the literature, which is abundant about the phenomena of soil–structure interaction and site effects, only a few examples highlight the effect of urban constructions on seismic motion. Jennings [JEN 70], Kanamori *et al.* [KAN 71], Guéguen *et al.* [GUÉ 00], Cornou *et al.* [COR 04], and Kim *et al.* [KIM 01] identified waves generated by the vibration of structures due to the inertial interaction of soil–structure. These waves are diffracted back into the soil and superimposed onto the incident seismic motion. In these studies, the structures were excited by internal or external forces applied at the top, and the wave diffused into soil was easily identifiable in the free-field recording station. This is not a local phenomenon, because Kanamori *et al.* [KAN 71], Favela *et al.* [FAV 02], and Kim *et al.* [KIM 01] observed it at distances greater than 50 km. For example, during the terrorist attacks on the twin towers of the World Trade Center, a seismological station managed by the Lamont-Doherty seismological network (Colombia University), situated about 40 kilometers from the impact site, not only showed the effect of the complete destruction of the towers, but also the effect of the vibration of the structures during the impact of the aircrafts (Figure 5.9).

Since then, a few observations have been made reutilizing the scheme originally considered by Guéguen *et al.* [GUÉ 00] at the Volvi test site in Greece, but applied to other soil–structure configurations [MUC 03, GAL 06], which confirm the existence of coupling between soil and structure motion. These observations have also been confirmed by simplified numerical modeling [WIR 96] and/or generalized to a group of buildings simulating a city [KHA 06, ISB 15], reproducing, in some cases, the monochromatic nature and duration of signals observed in Mexico City in 1985 during the Michoacan earthquake [AND 86]. This effect is not negligible and yields the question of how the response of the sites can be contaminated by the redistribution of seismic energy to the city surface.

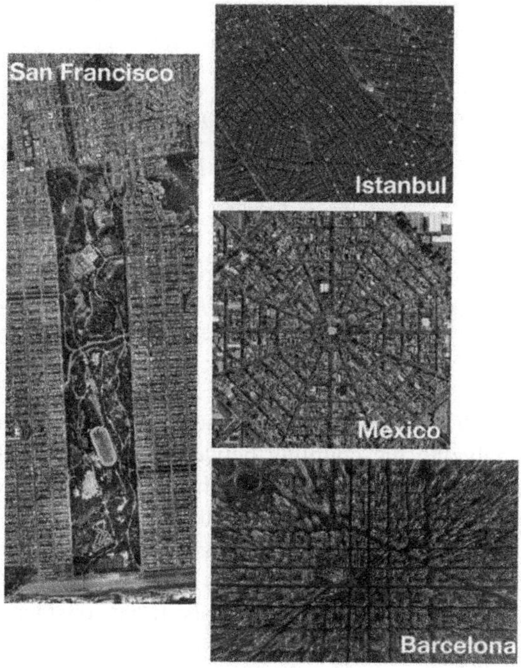

Figure 5.8. *Aerial images of cities exposed to seismic hazard and presenting a macroscopic organization of resonating elements. For a color version of this figure, see www.iste.co.uk/grange/soil.zip*

In Japan, as early as 1935, Sezawa and Kanai [SEZ 35] noted that the attenuation of earthquake effects with distance seemed more significant on the outskirts of cities, and they attributed this observation to the local coupling between the city and soil. Guéguen *et al.* [GUÉ 02] simulated the contribution of the Roma Norte district in Mexico City to the total seismic motion of soil. They showed that the evaluation of site effects in cities cannot be freed from integration into the process of the urban layer of buildings. They called this phenomenon the site–city interaction (SCI).

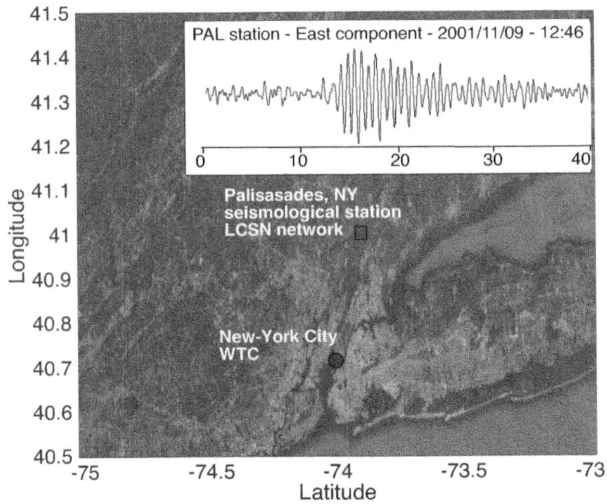

Figure 5.9. *Recording of soil motion 40 km from the WTC excited by the impact of aircrafts during the terrorist attacks of September 11, 2001, according to [KIM 01]. For a color version of this figure, see www.iste.co.uk/grange/soil.zip*

Wirgin and Bard [WIR 96] are certainly the first to have conducted numerical-based studies for analyzing the coupling between soil and structure motion when these structures are subjected to earthquakes. This idea was mainly motivated by the signature recordings of soil motion over the lake-bed area of Mexico City. In fact, as shown in Figure 5.10, the motion in the lake area certainly shows considerable amplitude amplification, due to the site effects of the surface soft layer, as well as remarkable lengthening and monochromatic beats. If we consider the 1D response of the Mexico City soil column, which has been well documented since the Michoacan earthquake of 1985, the surface motion can be calculated based on the motion recorded at the rock. It can be observed that the amplitude is well rendered, but not at all the monochromatic oscillations or the duration of the ground motion. On the contrary, in this area, there is a dense and massive habitat built on a flexible and superficial soil formation that favors the phenomena of soil–structure interaction. The contribution of structure motion to the observed soil motion was also encouraged by the similarity of the vibration frequencies of the structures and the lake surface layer, which meant that strong coupling had been established during the earthquake.

Guéguen *et al.* [GUÉ 02] demonstrated, with a simple analytical model, that in the time domain, motion of the wave field radiated by all structures was of the same order of magnitude as the incident seismic motion, thus demonstrating the importance of the site–city effect in the case of Mexico City. Despite some differences, mainly due to the simplicity of the model used, the calculated total field (direct + radiated by the structures) presents similarities with the records observed for the earthquake in question, particularly with regard to the lengthening and characteristic beats in the Mexico City area. This phenomenon does not seem to increase seismic motion, but rather perturb it with direct consequences for seismic risk. Indeed, Guéguen *et al.* [GUÉ 02] reached the conclusion that the estimation of site effects using the spectral ratio method [BOR 70] and observed on data could only be explained by integrating the total field for the calculation of site effects.

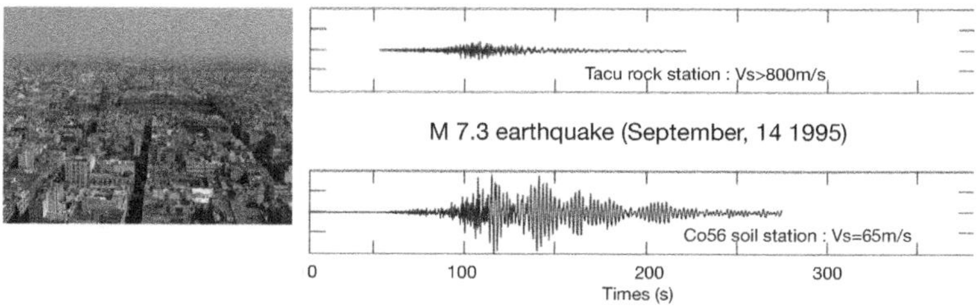

Figure 5.10. *Accelerometer recordings in a station at the rock (on top) and in the lake area of Mexico City (at the bottom) during a significant earthquake. For a color version of this figure, see www.iste.co.uk/grange/soil.zip*

An analysis of the energy ratios between the urban layer and the soil layer allowed Guéguen *et al.* [GUÉ 02] to evaluate urban configurations that could produce a site–city interaction phenomenon. The maximum kinetic energy Es of all the sediment filling and the maximal kinetic energy Eb of the n buildings are successively calculated. Guéguen *et al.* [GUÉ 02] thus propose the following simplified relation:

$$\frac{E_b}{E_s} = 2\sum_{i=1}^{n} \frac{S_{bi}}{S_s}\frac{H_{bi}}{H_s}\frac{f_s^2}{f_{bi}^2} \qquad [5.5]$$

which indicates that this phenomenon is essentially controlled by three terms that are:

– the urban density that represents the area of land occupied by buildings, expressed as the ratio between the total built surface area and the area of the site under consideration, i.e. $\frac{S_{bi}}{S_s}$;

– the mass contrast, and therefore of geometry, between the city and soil, characterized by the ratio between the average height of the buildings that compose the city and the thickness of the sedimentary filling, i.e. $\frac{H_{bi}}{H_s}$;

– the coincidence between the resonance frequencies of the soil layer and of the buildings, which favors the trapping of surface waves emitted by the vibration of buildings, namely $\frac{f_s^2}{f_{bi}^2}$.

Groby and Wirgin [GRO 04] also demonstrated, using modeling, that identical rigid blocks regularly distributed over half a space significantly altered the amplitude and duration of seismic motion. Kham *et al.* [KHA 06] observed a perturbation outside the city, exacerbated when the urban density is high and there is resonance between the city and the basin. They also digitally showed that:

– in a periodic city, the ISV effects are beneficial: the soil motion is reduced;

– this reduction increases with urban density and is maximal when construction and soil frequencies coincide. Under optimal conditions (the highest density and perfect resonance), the reduction reaches 50%;

– the density effect can, however, be significant even when the frequencies do not coincide. For example, the energy of the radiated field in relation to the incident field is of the same order (67%) in "low density – resonance" and "strong density – no resonance" situations;

– these reduction effects decrease significantly when the regularity of the building is broken. The reduction for non-periodic cities does not exceed 15%, which can be explained by the low number of buildings at 2 Hz and/or the low group effect due to the irregular layout.

Nevertheless, unlike Groby and Wirgin [GRO 04], Kham *et al.* [KHA 06] showed that the spatial consistency of seismic movement in the urban field was very perturbed and that the motion of structures and soil was reduced due to soil–structure–soil effects, and the greatest perturbations in soil motion related to the presence of structures appearing on the periphery of the city. This simulation resumes the observations made by Sezawa and Kanai as early as 1935 [SEZ 35].

At the scale of a city composed of a multitude of resonant structures, it is easy to imagine strong coupling between multiple structures and soil, redistributing the incident seismic energy. This raises questions about the validity of seismic evaluations of soil motion in urban areas, the understanding and interpretation of the site effects observed during earthquakes, and the way to design structures without integrating their urban environment in the process. This collective effect of coupled resonators has already been described in physics at the laboratory scale through the concept of locally resonant metamaterials, for example for a plate wave interacting with a dense collection of metal rods perpendicularly attached to it [RUP 14]. Applied to geophysics, where the earth replaces the plate for surface waves and a dense distribution of buildings would play the role of resonant metal rods, these metamaterials would cause a redistribution of energy of seismic waves [BRÛ 14, NIC 14, COL 16] and frequency band-gaps could appear. We might then ask to what extent the urban organization can contribute to the emergence of these band-gaps, in which the seismic energy is controlled and reduced, ultimately allowing seismic-resistant developments to be devised, i.e. structures that, organized according to a particular scheme, could create anti-seismic barriers.

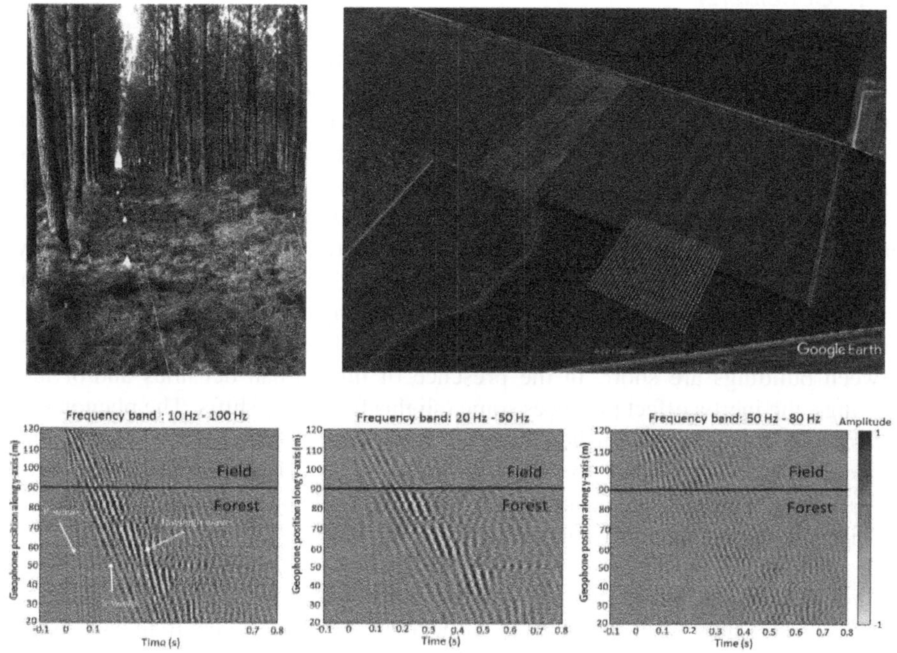

Figure 5.11. *Aerial and forest view of the deployment (yellow dots and yellow flags) of seismic sensors in METAFORET, at the interface between a field and a forest. The curves below represent the propagation of recorded waves on the vertical component of a line of seismic sensors during a test with a vibrator placed in the field. The black line represents the forest-field boundary. All three figures correspond to frequency bands applied by filtering to the recorded data. A clear transition can be seen around 50 Hz that corresponds to the vertical resonance frequency of the trees (according to [ROU 18]). For a color version of this figure, see www.iste.co.uk/grange/soil.zip*

We are still a long way from that, since the relationship between the vibration mode of the structures and the wavelengths involved is yet to be clarified. However, after laboratory experiments [RUP 14] and numerical modeling, a geophysical-scale experiment was conducted. It consisted of measuring the propagation of surface waves at the border between a field and a dense forest using a network of 1000 sensors [ROU 18]. The distance between the seismic sensors was equal to the distance between the trees (Figure 5.10). Early analyses clearly show prohibited frequency bands, with couplings between surface waves, also very present in sedimentary fillings of large exposed urban centers, and trees that resonate according to their vertical and horizontal mode. This dense and irregular forest is ultimately an analogue of a city, the concept of meta-forest can extend to that of meta-city, which would allow us to consider the control of seismic motion within the urban center and for certain frequency bands.

5.1.3. *Conclusion*

From this compilation of observations, specific experiments and numerical computations, we can draw some conclusions about the effects of constructions on seismic motion in urban areas. These results first indicate that the effect exists: soil motion is significantly perturbed in the immediate vicinity of a building and centrifuge trials indicate that the buildings "talk" to each other through soil. Computer simulations support the existence of strong interaction under favorable conditions (presence of resonance between soil and constructions and high urban density). The physical origin is multiple, mixing kinematic and inertial interactions: waves are generated in soil at the base of the vibrating buildings and they are trapped in the surface layers when the contrast is sufficient. Group effects also occur when distances between buildings are short. In the presence of high urban densities and/or large buildings, the inertia effect produces some (slight) frequency shifts. The phenomenon is particularly exacerbated when there is coincidence between soil and building frequencies: the ingredients for a significant SCI are therefore the simultaneous presence of a thin layer of soft sediments and dense urbanization with homogeneous buildings exhibiting similar frequencies.

In situations of strong multiple interaction, the effects appear to be generally beneficial: soil motion in urban areas is decreased for particular frequency bands. However, local amplifications cannot be completely eliminated. The overall beneficial effect should, nonetheless, be balanced by the fact that the SCI significantly increases the variability of seismic motion in urban areas. Some significant amplifications can locally occur, currently unpredictable because they are highly dependent on the incident wave field (frequency and phase). On the contrary, areas at the edges of dense and homogeneous urban centers are exposed to increased soil motion as a consequence of the effects of the SCI outside the urbanized area.

From a seismological point of view, it seems important to analyze seismic recordings taking into account the urban environment, in addition to the effects of sources, propagation and site effects. This issue is particularly important when dense networks are deployed in cities for seismic wave field analyses. The observation and analysis of the distribution of damage should also take into account the SCI as damage may not be due solely to vulnerability variability or site effects: from the transition from site–city interaction to the concept of meta-city.

From a seismic risk perspective, the main lesson to be drawn is that structural and site response may undergo modifications due to the urban environment. This could lead to many unpredictable developments, for example in urban planning (by trying to design "optimal land use" in order to reduce soil motion) and in the time evolution of risk (risk may change according to constructions and demolitions). Nevertheless, before confirming these consequences, the next necessary step is to obtain indisputable experimental evidence of the occurrence of these effects in actual cities subjected to earthquakes. It will therefore be necessary to specifically equip cities following particular instrumental methods and to use new sophisticated signal processing techniques to separate incident waves from the waves induced by the structures.

While there is still a need for additional 3D computations, these new results confirm that full-scale observation data in actual buildings are much more reliable than the information provided by sophisticated laboratory or modeling experiments, due to the *in situ* evaluation of the structure, its environment and boundary conditions. There are now signal processing algorithms that can provide new and valuable information about unexpected couplings or behaviors; these algorithms could be used to test and adjust new models and methods to improve structural health monitoring strategies.

This chapter was supported by funds from The METAFORET project funded by the French National Research Agency (ANR), the LabEx OSUG@2020 (Investissements d'avenir-ANR10LABX56), and the URBASIS-EU ITN project from the European Union's H2020 research and innovation program under the Marie Sklodowska-Curie Grant, Agreement Number 813137.

5.2. References

[AND 86] ANDERSON J.G., BODIN P., BRUNE J.N. *et al.*, "Strong ground motion from the Michoacan, Mexico, earthquake", *Science*, vol. 233, no. 4768, pp. 1043–1049, 1986.

[APS 87] APSEL R., LUCO J., "Impedance functions for foundations embedded in a layered medium: An integral equation approach", *Earthquake Engineering and Structural Dynamics*, vol. 15, pp. 213–231, 1987.

[AVI 96] AVILES J., PEREZ-ROCHA L., "Evaluation of interaction effects on the system period and the system damping due to foundation embedment and layer depth", *Soil Dynamics and Earthquake Engineering*, vol. 15, no. 1, pp. 11–27, 1996.

[BIE 75] BIELAK J., "Dynamic behavior of structures with embedded foundations", *Journal of Earthquake Engineering and Structural Dynamics*, vol. 3, no. 3, pp. 259–274, 1975.

[BOR 70] BORCHERDT R.D., "Effects of local geology on ground motion near San Francisco Bay", *Bulletin of the Seismological Society of America*, vol. 60, pp. 29–61, 1970.

[BRA 14] BRAY J., DASHTI S., "Liquefaction-induced building movements", *Bulletin of Earthquake Engineering*, vol. 12, no. 3, pp. 1129–1156, 2014.

[BRÛ 14] BRÛLÉ S., JAVELAUD E.H., ENOCH S. *et al.*, "Experiments on seismic metamaterials: Molding surface waves", *Physical Review Letters*, vol. 112, no. 3, pp. 133–901, 2014.

[CHA 90] CHAVEZ-GARCIA F., BARD P.-Y., "Surface ground motion modifications by the presence of a thin resistant layer. Applications to Mexico City", 9th *European Conference on Earthquake Engineering*, vol. 4-B, Moscow, pp. 37–46, 1990.

[CHA 03] CHAZELAS J., GUÉGUEN P., BARD P. *et al.*, "Modélisation de l'effet site-ville en modèles réduits centrifugés", *Actes du 6ème colloque National AFPS*, vol. I, Palaiseau, France, pp. 245–252, 2003.

[CHA 15] CHANDRA J., GUÉGUEN P., STEIDL J. *et al.*, "In-situ assessment of the G-i3 curve for characterizing the nonlinear response of soil: Application to the Garner Valley Downhole Array (GVDA) and the Wildlife Liquefaction Array (WLA)", *Bulletin of the Seismological Society of America*, vol. 105, no. 1A, pp. 993–1010, 2015.

[CHA 16] CHANDRA J., GUÉGUEN P., BONILLA F., "On the use of the seismic interferometry technique for testing PGV/Vs as a proxy for predicting nonlinear soil response", *Soil Dynamics and Earthquake Engineering*, vol. 85, pp. 146–160, 2016.

[CHA 19] CHANDRA J., GUÉGUEN P., "Nonlinear response of soil–structure systems using dynamic centrifuge experiments", *Journal of Earthquake Engineering*, vol. 23, no. 10, pp. 1719–1741, 2019.

[CLI 06] CLINTON J.F., BRADFORD S.C., HEATON T.H. *et al.*, "The observed wander of the natural frequencies in a structure", *Bulletin of the Seismological Society of America*, vol. 96, no. 1, pp. 237–257, 2006.

[COL 16] COLOMBI A., ROUX P., GUENNEAU S. *et al.*, "Forests as a natural seismic metamaterial: Rayleigh wave bandgaps induced by local resonances", *Scientific Reports*, vol. 6, no. 19238, 2016.

[COR 04] CORNOU C., GUÉGUEN P., BARD P.-Y. *et al.*, "Ambient noise energy bursts observation and modeling: Trapping of harmonic structure-soil induced-waves in a topmost sedimentary layer", *Journal of Seismology*, vol. 8, pp. 507–524, 2004.

[COU 14] COURBOULEX F., DELOUIS B., DUJARDIN A. *et al.*, "The two events of Barcelonette (French Alps), 2012 (MW 4.1) and 2014 (MW 4.9) - The role of directivity on ground motions, macroseismic intensities and site", *Second European Conference on Earthquake Engineering and Seismology*, Istanbul, Turkey, 25–29 August 2014.

[FAV 02] FAVELA J., HEATON T.H., TANIMOTO T., "Far-field energy radiation from a building under harmonic excitation", *AGU Fall Meeting Abstracts*, no. A1169, http://adsabs.harvard.edu/abs/2002AGUFM.S12A1169F (last accessed September 2016), 2002.

[GAL 06] GALLIPOLI M.R., MUCCIARELLI M., PONZO F. *et al.*, "Buildings as a seismic source: Analysis of a release test at Bagnoli", *Bulletin of the Seismological Society of America*, vol. 96, no. 6, pp. 2457–2464, 2006.

[GAZ 83] GAZETAS G., "Analysis of machine foundation vibrations: State of the art", *Soil Dynamics and Earthquake Engineering*, vol. 2, pp. 2–42, 1983.

[GAZ 91] GAZETAS G., "Foundation vibrations", in FANG H.-Y. (ed.), *Foundation Engineering Handbook*, 2nd edition, Van Nostrand Reinhold, New York, 1991.

[GHO 06] GHOSH B., MADABHUSHI S., "Centrifuge modeling of seismic soil structure interaction effects", *Nuclear Engineering Design*, vol. 237, no. 8, pp. 887–896, 2006.

[GRO 04] GROBY J.-P., WIRGIN A., "Seismic motion in urban sites consisting of blocks in welded contact with a soft layer overlying a hard half-space", *Geophysical Journal International*, vol. 172, no. 2, pp. 725–758, 2004.

[GUÉ 00] GUÉGUEN P., BARD P., SOUSA OLIVEIRA C., "Experimental and numerical analysis of soil motions caused by free vibrations of a building model", *Bulletin of the Seismological Society of America*, vol. 90, no. 6, pp. 1464–1479, 2000.

[GUÉ 02] GUÉGUEN P., BARD P.-Y., CHAVEZ-GARCIA F., "Site-city interaction in Mexico City-like environments: An analytical study", *Bulletin of the Seismological Society of America*, vol. 92, no. 2, pp. 794–811, 2002.

[GUÉ 05] GUÉGUEN P., BARD P.-Y., "Soil-structure and soil-structure-soil interaction: Experimental evidence at the Volvi test site", *Journal of Earthquake Engineering*, vol. 9, no. 5, pp. 657–693, 2005.

[GUÉ 07] GUÉGUEN P., CORNOU C., GARAMBOIS S. *et al.*, "On the limitation of the H/V spectral ratio using seismic noise as an exploration tool: Application to the Grenoble valley, a small apex ratio basin", *Pure and Applied Geophysics*, vol. 164, no. 1, pp. 1–20, 2007.

[GUÉ 16a] GUÉGUEN P., "Predicting nonlinear site response using spectral acceleration vs. PGV/Vs30: A case history using the Volvi-test site", *Pure and Applied Geophysics*, vol. 173, no. 6, pp. 2047–2063, 2016.

[GUÉ 16b] GUÉGUEN P., JOHNSON P., ROUX P., "Nonlinear dynamics induced in a structure by seismic and environmental loading", *The Journal of the Acoustical Society of America*, vol. 140, no. 1, pp. 582–590, 2016.

[GUÉ 17a] GUÉGUEN P., COLOMBI A., "Experimental and numerical evidence of the clustering effect of structures on their response during an earthquake: A case study of three identical towers in the city of Grenoble, France", *Bulletin of the Seismological Society of America*, vol. 106, no. 6, pp. 2855–2864, 2017.

[GUÉ 17b] GUÉGUEN P., LANGLAIS M., GARAMBOIS S. *et al.*, "How sensitive are site effects and building response to extreme cold temperature? The case of the Grenoble's (France) City Hall building", *Bulletin of Earthquake Engineering*, vol. 15, no. 3, pp. 889–906, 2017.

[GUY 99] GUYER R. A., JOHNSON P. A., "Nonlinear mesoscopic elasticity: Evidence for a new class of materials", *Physics Today*, vol. 52, pp. 30–6, 1999.

[HAD 93] HADJIAN A., "Seismic soil-structure interaction: A full circle", *Memorias X Congreso Nacional de Ingenieria Sismica*, Puerto Vallarta, Mexico, pp. 1–16, 1993.

[HIL 84] HILL N., LEVANDER A., "Resonances of low-velocity layers with lateral variations", *Bulletin of the Seismological Society of America*, vol. 74, pp. 521–537, 1984.

[HSI 62] HSIEH T., "Foundation vibrations", *Proceedings of Institution of Civil Engineers*, vol. 22, pp. 211–225, 1962.

[ISB 15] ISBILIROGLU Y., TABORDA R., BIELAK J., "Coupled soil-structure interaction effects of building clusters during earthquakes", *Earthquake Spectra*, vol. 31, no. 1, pp. 463–500, 2015.

[JOH 05] JOHNSON P., SUTIN A., "Slow dynamics and anomalous nonlinear fast dynamics in diverse solids", *The Journal of the Acoustical Society of America*, vol. 117, pp. 124–30, 2005.

[JEN 70] JENNINGS P.C., "Distant motion from a building vibration test", *Bulletin of the Seismological Society of America*, vol. 60, no. 6, pp. 2037–2043, 1970.

[KAN 71] KANAMORI H., MORI J., ANDERSON D. *et al.*, "Seismic excitation by the space shuttle Columbia", *Nature*, vol. 349, pp. 781–782, 1971.

[KAU 74] KAUSEL E., Forced vibrations of circular foundations on layered media, Research Report no. R77-3, MIT, Boston, USA, 1974.

[KAU 10] KAUSEL E., "Early history of soil–structure interaction", *Soil Dynamics and Earthquake Engineering*, vol. 30, no. 9, pp. 822–832, 2010.

[KHA 06] KHAM M., SEMBLAT J.-F., BARD P.-Y. *et al.*, "Seismic site–city interaction: Main governing phenomena through simplified numerical models", *Bulletin of the Seismological Society of America*, vol. 96, no. 5, pp. 1934–1951, 2006.

[KIM 01] KIM W.Y., SYKES L., ARMITAGE J. *et al.*, "Seismic waves generated by aircraft impacts and building collapses at World Trade Center, New-York City", *Eos, Transactions American Geophysical Union*, vol. 82–47, p. 565, 2001.

[KIT 99] KITADA Y., KINOSHITA M., IGUCHI M. *et al.*, "Soil-structure interaction effect on an NPP reactor building. Activities of Nupec: Achievements and the current status", in CELEBI M., OKAWA I. (eds), *Proc. UJNR Workshop on Soil-Structure Interaction*, Paper no. 18, Menlo Park, California, September 22–23, 1999.

[LAM 04] LAMB H., "On the propagation of tremors over the surface of an elastic solid", *Philosophical Transactions of the Royal Society*, vol. A-203, pp. 1–42, 1904.

[LEB 01] LEBRUN B., HATZFELD D., BARD P., "A site effect study in urban area: Experimental results in Grenoble (France)", *Pure and Applied Geophysics*, vol. 158, pp. 2543–2557, 2001.

[LEV 85] LEVANDER A.R., HILL N., "P-SV resonances in irregular low-velocity surface layers", *Bulletin of the Seismological Society of America*, vol. 75, pp. 847–864, 1985.

[LUC 71] LUCO J., WESTMANN R., "Dynamic response of circular footing", *Journal of Engineering Mechanics, ASCE*, vol. 97, no. EM 5, p. 1381, 1971.

[LUC 74] LUCO J., "Impedance functions for a rigid foundation on a layered medium", *Nuclear Engineering and Design*, vol. 31, pp. 204–217, 1974.

[LUC 88] LUCO J., TRIFUNAC M., WONG H., "Isolation of soil structure interaction effects by full scale forced vibration tests", *Earthquake Engineering and Structural Dynamics*, vol. 16, no. 1, pp. 1–21, 1988.

[LYS 65] LYSMER J., Vertical motions of rigid footings, PhD thesis, University of Michigan, 1965.

[MAS 13] MASON H., TROMBETTA N., CHEN Z. *et al.*, "Seismic soil-foundation-structure interaction observed in geotechnical centrifuge experiments", *Soil Dynamics and Earthquake Engineering*, vol. 48, pp. 162–174, 2013.

[MEH 07] MEHTA K., SNIEDER R., GRAIZER V., "Downhole receiver function: A case study", *Bulletin of the Seismological Society of America*, vol. 97, no. 5, pp. 1396–1403, 2007.

[MIC 10a] MICHEL C., GUÉGUEN P., "Time-frequency analysis of small frequency variations in civil engineering structures under weak and strong motion", *Structural Health Monitoring*, vol. 9, no. 2, pp. 159–171, 2010.

[MIC 10b] MICHEL C., GUEGUEN P., EL AREM S. *et al.*, "Full scale dynamic response of an RC building under weak seismic motions using earthquake loadings, ambient vibrations and modelling", *Earthquake Engineering and Structural Dynamics*, vol. 39, no. 4, pp. 419–441, 2010.

[MOS 87] MOSLEM K., TRIFUNAC M., "Spectral amplitudes of strong earthquake accelerations recorded in buildings", *Soil Dynamics and Earthquake Engineering*, vol. 6, no. 2, pp. 100–107, 1987.

[MUC 03] MUCCIARELLI M., GALLIPOLO M., PONZO F. *et al.*, "Seismic waves generated by oscillating buildings: Analysis of a release test", *Soil Dynamics and Earthquake Engineering*, vol. 23, pp. 255–262, 2003.

[NAK 12] NAKATA N., SNIEDER R., "Estimating near-surface wave velocities in Japan by applying seismic interferometry to KiK-net data", *Journal of Geophysical Research*, vol. 117, no. B01308, 2012.

[NIC 14] NICOLETTI O., "Seismic cloaks", *Nature Materials*, vol. 13, no. 5, pp. 428, 2014.

[PÉQ 08] PÉQUEGNAT C., GUÉGUEN P., HATZFELD D. *et al.*, "The French Accelerometric Network (RAP) and National Data Centre (RAP-NDC)", *Seismological Research Letters*, vol. 79, no. 1, pp. 79–89, 2008.

[PET 16] PETROVIC B., PAROLAI S., "Joint deconvolution of building and downhole strong-motion recordings: Evidence for the seismic wavefield being radiated back into the shallow geological layers", *Bulletin of the Seismological Society of America*, vol. 106, pp. 1720–1732, 2016.

[PRI 00] PRIESTLEY M. J.N., "Performance based seismic design", *Bulletin of the New Zealand Society for Earthquake Engineering*, vol. 33, no. 3, pp. 325–346, 2000.

[REI 36] REISSNER E., "Stationare, axialsymmetrische, durch eine shut-telnde masse erregte shwimgugen eines homogenen elastischen halbraumesn", *Ingenieur-Archiv*, vol. 7, p. 381, 1936.

[RIC 67] RICHART F., WHITMAN R., "Comparison of footing vibrations tests with theory", *Journal of the Soil Mechanics and Foundations Division*, vol. 93, pp. 143–168, 1967.

[ROU 18] ROUX P., BINDI D., BOXBERGER T. *et al.*, "Toward seismic metamaterials: The METAFORET project", *Seismological Research Letters*, vol. 89, pp. 582–593, 2018.

[RUP 14] RUPIN M., LEMOULT F., LEROSEY G. *et al.*, "Experimental demonstration of ordered and disordered multiresonant metamaterials for lamb waves", *Physical Review Letters*, vol. 112, no. 23, pp. 234–301, 2014.

[ŞAF 98] ŞAFAK E., "New approach to analyzing soil-building systems", *Soil Dynamics and Earthquake Engineering*, vol. 17, no. 7, pp. 509–517, 1998.

[ŞAF 99] ŞAFAK E., "Wave-propagation formulation of seismic response of multistory buildings", *Journal of Structural Engineering*, vol. 125, no. 4, pp. 426–437, 1999.

[SEM 98] SEMBLAT J., LUONG M., "Wave propagation through soils in centrifuge testing", *Journal of Earthquake Engineering*, vol. 2, no. 1, pp. 147–171, 1998.

[SEZ 35] SEZAWA K., KANAI K., "Decay in seismic vibration of a simple or tall building structure by dissipation of their energy into the ground", *Bulletin of the Earthquake Research Institute*, vol. XIII, no. 3, pp. 681–697, 1935.

[SNI 06] SNIEDER R., SAFAK E., "Extracting the building response using seismic interferometry: Theory and application to the Millikan Library in Pasadena, California", *Bulletin of the Seismological Society of America*, vol. 96, no. 2, pp. 586–598, 2006.

[STE 98] STEWART J., FENVES G.L., "System identification for evaluating soil–structure interaction effects in buildings from strong motion recordings", *Earthquake Engineering and Structural Dynamics*, vol. 27, no. 8, pp. 869–885, 1998.

[STE 99] STEWART J., FENVES G., SEED R., "Seismic soil-structure interaction in buildings. I: Analytical methods", *Journal of Geotechnical and Geoenvironmental Engineering, ASCE*, vol. 125, no. 1, pp. 26–37, 1999.

[TOD 08] TODOROVSKA M.I., TRIFUNAC M.D., "Impulse response analysis of the Van Nuys 7-storey hotel during 11 earthquakes and earthquake damage detection", *Structural Control and Health Monitoring*, vol. 15, no. 1, pp. 90–116, 2008.

[TRI 72] TRIFUNAC M., "Interaction of a shear wall with the soil for incident plane SH waves", *Bulletin of the Seismological Society of America*, vol. 62, pp. 63–83, 1972.

[TRI 10] TRIFUNAC M.D., TODOROVSKA M.I., MANIĆ M.I., BULAJIĆ B.Ä., "Variability of the fixed base and soil structure system frequencies of a building. The case of Borik 2 building", *Structural Control and Health Monitoring*, vol. 17, no. 2, pp. 120–151, 2010.

[TRO 13] TROMBETTA N., MASON H., CHEN Z. *et al.*, "Nonlinear dynamic foundation and frame structure response observed in geotechnical centrifuge experiments", *Soil Dynamics and Earthquake Engineering*, vol. 50, pp. 117–133, 2013.

[TRO 14] TROMBETTA N., MASON H., CHEN Z. *et al.*, "Nonlinear soil-foundation-structure and structure-soil-structure interaction: Centrifuge test observations", *Journal of Geotechnical and Geoenvironmental Engineering*, vol. 140, no. 5, pp. 04013057-(1-11), 2014.

[VEL 71] VELETSOS A., WEI Y., "Lateral and rocking vibration footings", *Journal of the Soil Mechanics and Foundations Division, ASCE*, vol. 97, pp. 1227–1248, 1971.

[WIR 89] WIRGIN A., "Amplification résonante du tremblement d'une chaîne de montagne cylindrique soumise à une onde SH", *Compte rendu de l'Académie des Sciences*, vol. 311, pp. 651–655, 1989.

[WIR 96] WIRGIN A., BARD P.-Y., "Effects of buildings on the duration and amplitude of ground motion in Mexico City", *Bulletin of the Seismological Society of America*, vol. 86, pp. 914–920, 1996.

[WOL 85] WOLF J., *Dynamic Soil-Structure Interaction*, Prentice Hall, Englewood Cliffs, New Jersey, 1985.

[WON 77] WONG H., TRIFUNAC M., WESTERMO B., "Effects of surface and subsurface irregularities on the amplitudes of monochromatic waves", *Bulletin of the Seismological Society of America*, vol. 67, pp. 353–368, 1977.

List of Authors

Benjamin CERFONTAINE
School of Science and Engineering
University of Dundee
UK

Robert CHARLIER
Dpt. ArGEnCo
University of Liège
Belgium

Andrea COLOMBI
Institute of Structural Engineering
Swiss Federal Institute of Technology
Zurich
Switzerland

Daniel DIAS
3SR Laboratory
Grenoble Alpes University
France

Stéphane GRANGE
GEOMAS Laboratory
INSA
Lyon
France

Philippe GUEGUEN
Earth Science Institute (ISTerre)
Grenoble Alpes University
CNRS
IFSTTAR
France

Orianne JENCK
3SR Laboratory
Grenoble Alpes University
France

Panagiotis KOTRONIS
GeM
Centrale Nantes
France

Philippe ROUX
Earth Science Institute (ISTerre)
Grenoble Alpes University
CNRS
IFSTTAR
France

Diana SALCIARINI
Department of Civil and Environmental Engineering
University of Perugia
Italy

Claudio TAMAGNINI
Department of Civil and
Environmental Engineering
University of Perugia
Italy

Pascal VILLARD
3SR Laboratory
Grenoble Alpes University
France

Index

C, D, F

contact mechanics, 48
cyclic loading, 56, 63, 71–73
discrete element method, 55–57, 59, 61, 62, 68–71, 76, 80, 96
dynamic loading, 101, 105, 112, 120, 122, 137
finite
  difference method, 101, 102, 104, 112, 117, 122, 126, 137
  difference model, 119, 133
  element modeling, 2

G, H, I

geosynthetic, 56, 76–96
geotechnical
  engineering, 101, 102, 105, 108, 112, 138
  structure, 55, 56, 61, 76
hydro-mechanical couplings, 1, 2, 6, 30, 37, 43, 48
hypoplasticity, 145, 149, 151–154, 156, 158–160, 162, 165, 167, 171–173
interface, 1–3, 5–11, 13–20, 22–24, 26–30, 32–35, 38, 43, 46–49, 56, 59, 60, 63, 65, 70, 72, 77, 84, 88–90, 96

L, M, N

lateral loading, 1, 22, 25, 37, 40–42, 47, 48
macroelement, 143–145, 148–151, 153, 154, 156, 158–160, 162–165, 167, 169–174
metamaterials, 183, 200
nonlinear behavior, 143–145, 147, 149, 151, 159, 161, 162
numerical modeling, 101, 109, 111, 117, 122, 124, 126, 127, 129, 132, 133, 135, 136, 138

O, P, R, S

offshore foundations, 1, 20, 34, 49
partially drained behavior, 2, 7, 48
piles, 56, 61–68, 70–73, 75, 76
  vertical rigid, 80, 121
plasticity, 144, 147, 151, 172
reinforcement, 55, 56, 59–62, 76, 77, 84, 87, 89, 96
retaining walls, 102, 106, 107, 110, 112
seismic risk, 199, 202
shallow foundation, 144, 146–148, 154, 160, 172
site–city interaction (SCI), 197, 199, 202

soil improvement, 102, 121, 129
soil–structure interaction (SSI), 101–105, 107, 108, 111–113, 115, 116, 119, 120, 123, 138
static loading, 102, 119
structural dynamic, 143
suction
  anchors, 2, 20, 21, 37, 49
    caisson, 1, 2, 7, 20–22, 24, 34, 48

T, U, W

tunnels, 102, 106, 109–112, 115–121
urban seismology, 181, 182, 183
waves, 181, 182, 185–187, 189, 190, 193, 196, 198–203

Other titles from

ISTE

in

Numerical Methods in Engineering

2021

GENTIL Christian, GOUATY Gilles, SOKOLOV Dmitry
Geometric Modeling of Fractal Forms for CAD
(Geometric Modeling and Applications Set – Volume 5)

2020

GEORGE Paul Louis, ALAUZET Frédéric, LOSEILLE Adrien, MARÉCHAL Loïc
Meshing, Geometric Modeling and Numerical Simulation 3: Storage, Visualization and In Memory Strategies
(Geometric Modeling and Applications Set – Volume 4)

SIGRIST Jean-François
Numerical Simulation, An Art of Prediction 2: Examples

2019

DA Daicong
Topology Optimization Design of Heterogeneous Materials and Structures

GEORGE Paul Louis, BOROUCHAKI Houman, ALAUZET Frédéric, LAUG Patrick, LOSEILLE Adrien, MARÉCHAL Loïc
Meshing, Geometric Modeling and Numerical Simulation 2: Metrics, Meshes and Mesh Adaptation
(Geometric Modeling and Applications Set – Volume 2)

MARI Jean-Luc, HÉTROY-WHEELER Franck, SUBSOL Gérard
Geometric and Topological Mesh Feature Extraction for 3D Shape Analysis
(Geometric Modeling and Applications Set – Volume 3)

SIGRIST Jean-François
Numerical Simulation, An Art of Prediction 1: Theory

2017

BOROUCHAKI Houman, GEORGE Paul Louis
Meshing, Geometric Modeling and Numerical Simulation 1: Form Functions, Triangulations and Geometric Modeling
(Geometric Modeling and Applications Set – Volume 1)

2016

KERN Michel
Numerical Methods for Inverse Problems

ZHANG Weihong, WAN Min
Milling Simulation: Metal Milling Mechanics, Dynamics and Clamping Principles

2015

ANDRÉ Damien, CHARLES Jean-Luc, IORDANOFF Ivan
3D Discrete Element Workbench for Highly Dynamic Thermo-mechanical Analysis
(Discrete Element Model and Simulation of Continuous Materials Behavior Set – Volume 3)

JEBAHI Mohamed, ANDRÉ Damien, TERREROS Inigo, IORDANOFF Ivan
Discrete Element Method to Model 3D Continuous Materials
(Discrete Element Model and Simulation of Continuous Materials Behavior Set – Volume 1)

JEBAHI Mohamed, DAU Frédéric, CHARLES Jean-Luc, IORDANOFF Ivan
Discrete-continuum Coupling Method to Simulate Highly Dynamic Multi-scale Problems: Simulation of Laser-induced Damage in Silica Glass (Discrete Element Model and Simulation of Continuous Materials Behavior Set – Volume 2)

SOUZA DE CURSI Eduardo
Variational Methods for Engineers with Matlab®

2014

BECKERS Benoit, BECKERS Pierre
Reconciliation of Geometry and Perception in Radiation Physics

BERGHEAU Jean-Michel
Thermomechanical Industrial Processes: Modeling and Numerical Simulation

BONNEAU Dominique, FATU Aurelian, SOUCHET Dominique
Hydrodynamic Bearings – Volume 1
Mixed Lubrication in Hydrodynamic Bearings – Volume 2
Thermo-hydrodynamic Lubrication in Hydrodynamic Bearings – Volume 3
Internal Combustion Engine Bearings Lubrication in Hydrodynamic Bearings – Volume 4

DESCAMPS Benoît
Computational Design of Lightweight Structures: Form Finding and Optimization

2013

YASTREBOV Vladislav A.
Numerical Methods in Contact Mechanics

2012

DHATT Gouri, LEFRANÇOIS Emmanuel, TOUZOT Gilbert
Finite Element Method

SAGUET Pierre
Numerical Analysis in Electromagnetics

SAANOUNI Khemais
Damage Mechanics in Metal Forming: Advanced Modeling and Numerical Simulation

2011

CHINESTA Francisco, CESCOTTO Serge, CUETO Elias, LORONG Philippe
Natural Element Method for the Simulation of Structures and Processes

DAVIM Paulo J.
Finite Element Method in Manufacturing Processes

POMMIER Sylvie, GRAVOUIL Anthony, MOËS Nicolas, COMBESCURE Alain
Extended Finite Element Method for Crack Propagation

2010

SOUZA DE CURSI Eduardo, SAMPAIO Rubens
Modeling and Convexity

2008

BERGHEAU Jean-Michel, FORTUNIER Roland
Finite Element Simulation of Heat Transfer

EYMARD Robert
Finite Volumes for Complex Applications V: Problems and Perspectives

FREY Pascal, GEORGE Paul Louis
Mesh Generation: Application to finite elements – 2nd edition

GAY Daniel, GAMBELIN Jacques
Modeling and Dimensioning of Structures

MEUNIER Gérard
The Finite Element Method for Electromagnetic Modeling

2005

BENKHALDOUN Fayssal, OUAZAR Driss, RAGHAY Said
Finite Volumes for Complex Applications IV: Problems and Perspectives